The Ethics of Artificial Intelligence in Defence

The Ethics of Artificial Intelligence in Defence

MARIAROSARIA TADDEO

OXFORD
UNIVERSITY PRESS

OXFORD
UNIVERSITY PRESS

Oxford University Press is a department of the University of Oxford. It furthers the University's objective of excellence in research, scholarship, and education by publishing worldwide. Oxford is a registered trade mark of Oxford University Press in the UK and certain other countries.

Published in the United States of America by Oxford University Press
198 Madison Avenue, New York, NY 10016, United States of America.

Library of Congress Cataloging-in-Publication Data
Names: Taddeo, Mariarosaria, author.
Title: The ethics of artificial intelligence in defence / Mariarosaria Taddeo.
Other titles: Ethics of artificial intelligence in defense
Description: New York, NY : Oxford University Press, [2024] |
Includes bibliographical references and index.
Identifiers: LCCN 2024023301 | ISBN 9780197745441 (hardback) |
ISBN 9780197745465 (epub)
Subjects: LCSH: Artificial intelligence—Military applications. |
Artificial intelligence—Moral and ethical aspects. |
Autonomous weapons systems—Moral and ethical aspects.
Classification: LCC UG479 .T35 2024 | DDC 172/.42—dc23/eng/20240729
LC record available at https://lccn.loc.gov/2024023301

DOI: 10.1093/oso/9780197745441.001.0001

The manufacturer's authorised representative in the EU for product safety is Oxford University Press España S.A. of El Parque Empresarial San Fernando de Henares, Avenida de Castilla, 2 – 28830 Madrid (www.oup.es/en or product.safety@oup.com). OUP España S.A. also acts as importer into Spain of products made by the manufacturer.

Contents

Preface

> Tell him he's wrong. War doesn't negate decency. It demands it, even more than in times of peace.
> —Khaled Hosseini, *The Kite Runner*

The foundations of many of the ideas presented in this book were laid nearly 15 years ago when I initiated a research project exploring the ethics of information warfare.[1] However, the concept for this book emerged only more recently, influenced by my involvement in discussions with various defence organisations of the ethics surrounding artificial intelligence (AI). During this collaboration, I observed a growing need for a unified framework that can identify, analyse, and address systematically the ethical challenges arising from the many current—and emerging—uses of AI in the defence sector, as well as guide their governance. With this book, I am to fulfil this need.

Clausewitz stated that "war is an act of force to compel our enemy to do our own will. Force, to counter opposite force, equips itself with the inventions of art and science" (Clausewitz 2008, 75). I believe this is too simplistic a view of the role that technologies play in the waging of war. Technologies are not mere tools to optimise the use of force; they also act as disrupting factors in our conceptualisation of war and of its ethical and legal implications. Consider, for example, nuclear technology and nuclear weapons. As weapons of mass destruction, they overrode

---

[1] Some of the chapters of this book were first planned as articles or conference papers. The full list of these publications can be found at the end of the acknowledgments.

the distinction between combatant and non-combatant, and in so doing overturned an idea of war in which discrimination was central, and upon which Just War Theory had relied until then. Similarly, the use of drones has contributed to a redefinition of our understanding of war as "a duel on a large scale" (Clausewitz 2008, 75), by exploding the idea of a symmetric confrontation where both sides take similar risks (Steinhoff 2013; Braun and Brunstetter 2013; Strawser 2013; Schulzke 2016). This is why I agree with Clarke that "when we apply ethics to war, we are left to shoot at a constantly moving target, and the ethics have to track that evolution" (2015, 19). The adoption of digital technologies in defence moved our target once again, with AI providing one of the starkest examples of how technologies redefine our understanding of relevant concepts (Floridi 2014). Consider, for example, our understanding of war as a coercive behaviours coupled to the use of force. For centuries, indeed since Cicero (106–43 BCE), we have regulated war by regulating the use of force (Cicero 2008). Fast-forward to 2014—when NATO[2] declared cyberspace to be a domain of warfare—and we see that this normative approach fails to work, given that war in cyberspace decouples coercive behaviour from the use of force. As a result, we find state actors engaging in non-kinetic, adversarial activities—*cyber warfare*—without adequate regulation or provision to limit the risks of escalation of conflicts, breaches of individual rights, and unnecessary threats to civilian infrastructures. Unfortunately, kinetic warfare still occurs, but cyber warfare has been on the rise in recent decades, posing pressing questions as to how to regulate state-run or state-sponsored non-kinetic cyber warfare.

The redefinition of our understanding of war continues with the adoption of AI in defence. Consider Clausewitz's famous trinity, which describes war as "composed of primordial *violence* . . . ; of

---

[2] *Wales Summit Declaration*, NATO website, published 5 September 2014, https://www.nato.int/cps/en/natohq/official_texts_112964.htm (accessed August 2024)

the play of *chance* and probability within which the creative spirit is free to roam; and of its element of subordination, as an instrument of policy, which makes it subject to *reason* alone" (Clausewitz 2008, 89). The use of AI has an impact on the violent element of the trinity. AI may remove violence (force) from warfare, for example by improving the efficacy of non-kinetic, cyber operations and making kinetic ones less necessary. At the same time, AI may change the way violence is exerted by removing human agents from the process of applying force, if/when fully autonomous weapon systems are deployed. AI may support human reasoning by improving situational awareness and supporting the decision-making process. In doing so it may de-anthropomorphise war waging and hinder the "creative spirit" to which Clausewitz refers, if human agents accept the recommendations of an AI system in an uncritical way.

AI is reshaping the way wars are waged but also the way in which these forces function, make decisions, obtain data and information, and operate in the different domains of war. Here, I consider the *entire* spectrum of changes, and their conceptual and ethical implications. This is why this is a book on the ethics of AI in defence—and not on the ethics of AI in war.

I make no claim as to whether we should consider the use of digital technologies and AI in defence as a revolution in military affairs—limited to mostly operational matters—or as the sparkle of military revolution. That is the idea that the changes in military affairs, strategies, tactics, and the conduct of warfare caused by the adoption of digital technologies will result in a shift in the balance of power and the way conflicts are resolved. Instead, my focus is entirely on the conceptual and ethical implications of this digitalisation and how to manage the related ethical risks and opportunities; whether the digitalisation of defence falls under the rubric of one case or the other labels or neither.

The emphasis on both conceptual and normative implications of AI in defence may safeguard this book against the risk of a swift obsolescence. The role of AI in defence is a pertinent subject, but one that can swiftly shift from being relevant to merely trendy if its analysis leans too heavily on descriptive or anecdotal accounts. The durability of this book will hinge on its ability to delve into the conceptual and normative aspects of a timely phenomenon, namely the digital revolution, to identify its ramifications for the foreseeable future, while avoiding becoming overly preoccupied with chronicling specific events.

My goal in this book is to provide an ethical framework to identify, analyse, and address systematically the ethical implications posed by the different uses of AI in the defence domain and to shape their governance. I do so by offering an analysis of the relevant conceptual and ethical challenges and outlining a normative framework to address them to guide decision-makers and policymakers in developing an ethical governance of AI in defence. To achieve this goal, I focus on the autonomy and learning capabilities of AI technologies and the level of agency that they enable. I then build on AI ethics and Just War Theory.

A key contribution of this book is the translation of the conceptual and normative analysis into recommendations for the governance of AI in defence. In this sense, the book reflects the idea that the ethics of AI works best as a *translational ethics* (Taddeo and Floridi 2018b, 752). The translational approach calls for as clear an analysis as possible; indeed there is no value in an obscure translation. So I have made every effort to maintain simplicity in both language and analysis. Nevertheless, I must acknowledge that this is not an introductory text on ethics of defence, military ethics or AI ethics. This book is intended for an audience with some level of expertise in AI ethics and Just War Theory.

The proposed analysis is structured around three categories of use of AI (Taddeo et al. 2021b): sustainment and support; adversarial and non-kinetic; and adversarial and kinetic. 'Sustainment and support uses' refers to the deployment of AI to support back-office functions, logistics, and uses of AI that aim to improve the security of the infrastructures and communication systems that underpin national defence services. This category of use also includes the use of AI to support strategic decision-making and wargaming practices. 'Adversarial and non-kinetic uses' refers to the deployment of AI for defensive and offensive cyber operations with non-kinetic aims. 'Adversarial and kinetic uses' refers to the integration of AI systems in combat operations. They range from the use of AI systems for threat identification to their use in lethal autonomous weapons systems.

In the rest of this book, chapter 1 provides the groundwork for the rest of the book and outlines the methodology and scope of the analysis. Chapter 2 offers five fundamental principles of AI ethics in defence and a methodology to implement them in a defence organisation. The following six chapters then focus on different uses of AI in defence, ranging from AI-augmented intelligence analysis to cyber warfare and cyber deterrence, to autonomous weapon systems. Readers interested in specific uses of AI in defence can skip between these latter chapters, provided they have read Chapters 1 and 2.

Before concluding this section, I shall address a question that may have arisen on reading the title of this book: "why bother with ethics of AI in defence?" I have come across this question many times when discussing this work with scholars, policymakers, and defence experts. I hope that the analysis of the ethical problems and opportunities associated with the use of AI in defence that I propose here provides sufficient evidence to justify deep thought and discussion on the ethics of AI in defence. However, as I do not engage with this question in the rest of the book, I shall take the opportunity to address it here.

The question is commonly asked by people belonging to two camps. The first one is the camp of those pursuing the use of AI in defence, who perceive ethics as a hindrance to the adoption of this technology, and which may reduce the advantage over our opponents who might acquire AI capabilities and use them without first considering the related ethical implications. This camp often embraces a realistic view of international politics. Pacifism motivates those belonging to the opposite camp. Here the assumption is that no good can come from defence and, *a fortiori*, from the use of AI in defence, hence there is no ethical reflection necessary, as any use of this technology in defence should be forbidden. According to this view, this type of ethical analysis is unnecessary and problematic, as it may end up legitimising the use of AI in defence.

I believe that the assumptions guiding both camps are untenable. On the one side, liberal democratic societies engage in a *just* war, that is, to defend themselves. It is unthinkable to engage in conflict to defend these societies without a regard for ethics—and thus the very fundamental values and rights that they aim to defend. This would be for liberal democracies to defeat themsleves. Realism is a dangerous view because it lets the opponent set the conditions under which a liberal democratic society respects and enacts its own fundamental values and rights. On the other side, pacifist views remain too idealistic for me. I agree that war is an absolute evil, but it is one with which humans have engaged—and continue to engage—routinely over the course of history. While aggression remains unjustifiable, the right of a state to defend itself and resist an aggressor also remains undeniable. Serious consideration of the ethics of war is crucial to avoid such a right leading to perfidy in war (Lippert-Rasmussen 2014), and hence to atrocities.

The reader who does not agree with these (political) reasons in support of the urgent need for an ethics of AI in defence may appreciate the view that Walzer proposes when he links the ethics of war to strategy. According to Walzer, both provide a language of justification; that is, ethical reasoning and strategic reasoning are both concerned with identifying risks and opportunities and defining a course of actions to mitigate the former and leverage the latter. In Walzer's words (2006, 15–16)

> the moral theorist . . . must come to grips with the fact that his rules are often violated or ignored—and with the deeper realisation that, to men at war, the rules often don't seem relevant to the extremity of their situation. But however he does this, he does not surrender his sense of war as a human action, purposive and premeditated, for whose effects someone is responsible. Confronted with the many crimes committed in the course of war, or with the crime of aggressive war itself, he searches for human agents. Nor is he alone in this search. It is one of the most important features of war, distinguishing it from the other scourges of mankind, that the men and women caught up in it are not only victims, they are also participants. All of us are inclined to hold them responsible for what they do. . . . [W]e may say of a general who experienced no difficulty of making a (really) difficult decision that he did not understand the strategic realities of his own position or that he was reckless and insensible of danger. And we might go on to argue, in the case of the general, that such a man has no business in fighting or leading others in battle, that he ought to . . . worry about the danger and take steps to avoid it. *Once again, the case is the same with moral decisions: soldier and statesmen ought to know the dangers of cruelty and injustice and worry about them and take steps to avoid them.* (Emphasis added)

This is no less true when considering the use of AI in defence and the risks that humans may rely on these technologies to perpetrate cruelty and injustice in war waging. The history of war reminds us of the ever-absurd level of atrocity that we can inflict on one another. When considering that AI is designed to maximise (even exceed) human performance, then the relevancy and need for ethical analyses informing the use of AI in defence should be self-evident.

Acknowledgements

This book could not have been completed without the support of my colleagues, friends, and family. I am lucky enough to bear a huge debt of gratitude with all of them. I shall begin with the colleagues with whom I have worked most closely over the years. I am deeply grateful to them for sharing with me their insights, intuitions, questions, and enthusiasm for understanding the digital revolution and to help steer it, even if in some small way, with our work. Rachel Azafrani, Alexander Blanchard, Josh Cowls, Prathm Juneja, Jakob Mökander, Jess Morley, Carl Öhman, Huw Roberts, David Sutcliffe, Christopher Thomas, Andreas Tsamados, Vincent Wang, David Watson, and Marta Ziosi have all contributed to refining my ideas about the ethics of AI in defence in significant ways.

The content of this book has been discussed with many colleagues in academia, government, and the defence establishment. I am grateful for all the suggestions, case studies, criticisms, and references that colleagues have shared with me over the years, some of which have helped me refine key parts of my work. I am particularly grateful to Al Banks, David McNeish, and Leila Kleineidam for their commitment to discussing AI ethics in the defence domain, and the time they spent reading and commenting on parts of my work. I also would like to acknowledge the support, suggestions, and words of encouragement that Rebecca Eynon, George Lucas, Ties Nijssen, and Selmer Bringsjord have given me over these years. They helped me navigate the moments of uncertainty that inevitably arose during this project. I am also grateful to Peter Ohlin for believing in it enough to publish it.

Parts of this book were first planned as research articles or book chapters (as indicated in the list of references) as part of different research projects I have led over the years. I am thankful to the funders for their support—particularly the Marie Skłodowska-Curie Actions, the John Fell Fund of the University of Oxford, the NATO Centre of Excellence on Cooperative Cyber Defence, and the Alan Turing Institute in London, the Defence Science and Technology Laboratory of the UK Ministry of Defence. Some of this funding was awarded with admirable foresight, way before the ethics of AI, particularly in defence, was an established area of research. I quickly pass over all the proposals that were not funded and the articles that were rejected. In an unenthusiastic way, I am grateful also for these failures. Each time, they enabled me to improve my project and sharpen my ideas.

I must acknowledge the unwavering support and kindness I have received from a remarkable group of friends: Antonella Giglio, Dominik Aschbrenner, Elena Brenna, Elisa Sacchi, Francesco Fermani, and Giorgia Brambilla Pisoni. To be able to count on their affection, to find in them supportive and honest sparring partners in disentangling my thoughts, break the academic bubble, and create great memories, has been crucial to maintaining my sanity while writing this book. I am also thankful to my family. From them I learned determination and stubbornness, which have been crucial to finishing this book. I have in them a source of pragmatic, at times tough, support, which keeps me grounded. My deep and enduring gratitude extends to Luciano Floridi and Kia Nobre. Finding precise words to express my affection and appreciation for them is challenging, for they are truly inspirational figures. Their work, generosity, integrity, sense of humour, boundless energy, brilliance, friendship, affection, support, and, above all, genuine enthusiasm for research serve as a constant source of motivation for me. I am only embarrassed that as good as their example has and continues to be, I am able to follow it only in part.

Despite all the extensive help I received, I am aware that there may still be some shortcomings and errors within the following pages; these are my sole responsibility.

The chapters of this book are adapted from the following publications:

Chapter 1: Taddeo et al. 2022; Taddeo, McNeish et al. 2021
Chapter 2: Taddeo et al. 2021b; Taddeo, Blanchard, and Thomas 2023
Chapter 3: Blanchard and Taddeo 2023
Chapter 4: Taddeo 2014a, 2016b, 2017a, 2018
Chapter 5: Taddeo 2017b, 2018a, 2018c
Chapter 6: Blanchard and Taddeo 2022b
Chapter 7: Taddeo and Blanchard 2022
Chapter 8: Blanchard and Taddeo 2022a, 2022b, 2022c

Most-Used Abbreviations

AI	artificial intelligence
AIA	AI for intelligence analysis
AWS	autonomous weapon systems
DL	deep learning
DoD	Department of Defence
EB	ethics board
HMT	human-machine teaming
ICRC	International Committee of the Red Cross
IHL	international humanitarian law
LAWS	lethal autonomous weapon systems
LLM	large language model
ML	machine learning
MoD	Ministry of Defence

1

The Groundwork for an Ethics of Artificial Intelligence in Defence

1. Introduction

The uses of AI in defence are various—from the use of machine-learning (ML) algorithm optimisation and predictive analysis to improve management of supplies and equipment, to analysis of vast amounts of data, to supporting decision-making for the application of force, and also in the use of force itself. From tweets to tanks, AI is now a key capability in defence and therefore the object of a global race for its development and use (Taddeo and Floridi 2018a). For example, since 2014, the Russian National Defense Control Center has been using AI to detect online threats. In 2017 the Chinese government issued its Next Generation AI Development Plan, of which military implementation of AI on the battlefield and in cyberspace is a crucial part. The UK and the US, France and Australia have published national defence strategies centred on the use of AI. NATO has been working to boost defence innovation, and AI has a key place in its strategy. AI is also being deployed in conflicts. In 2023, Isreal used AI systems for human target identification in Gaza, reportedly,[1] and AI is widely used in the war in Ukraine.[2]

---

[1] "'Lavender': The AI machine directing Israel's bombing spree" in Gaza +972 Magazine, published on 3 April 2024, https://www.972mag.com/lavender-ai-israeli-army-gaza/ (accessed August 3, 2023).

[2] "Roles and Implications of AI in the Russian-Ukrainian Conflict", Centre of New America Security website, published on 20 July 2023, https://www.cnas.org/publicati

The Ethics of Artificial Intelligence in Defence. Mariarosaria Taddeo, Oxford University Press.
© Oxford University Press 2024. DOI: 10.1093/oso/9780197745441.003.0001

The use of AI for national defence poses important ethical problems that combine ethical risks relating to the use of AI itself—for example, enabling human wrongdoing, reducing human control, removing human responsibility, devaluing human skills, and eroding human self-determination (Yang et al. 2018)—with those that follow from the use of force in warfare, like respecting of human dignity and the risk of breaching the principles of Just War Theory.

Because of the range of possible applications and the wide and complex set of ethical risks that need to be addressed, it is difficult to develop a coherent and systemic ethical analysis of AI in defence. Problems begin with the definition of AI (because this technology draws on such a variety of approaches, methods, and models) and continue when we try to determine the scope of the ethical analysis. To address this problem, we may think of developing a taxonomy of ethical issues of AI in defence, but this is unfeasible and of little value: the taxonomy would be quickly outdated by the rapid developments in AI technologies and their application to new uses. At the same time, different ethical problems may become evident only when considering AI from different points of view. For example, some ethical problems of AI are inherent to the design and development process, while others emerge within specific domains and purposes of deployment. This makes it difficult to identify the right level of analysis. Analyses that disregard the characteristics of the defence domain and purposes of deployment risk being too generic to provide any concrete guidance. At the same time, ethical analyses that limit themselves to use of AI in a specific domain still need to be harmonised with the broader set of values underpinning our societies. To avoid these limitations, or at the very least to be aware of them, assumptions concerning the scope of the analysis

ons/commentary/roles-and-implications-of-ai-in-the-russian-ukrainian-conflict (accessed August 3, 2024) and its methodology need to be clarified up front. This is the goal of this chapter.

Here, I provide the groundwork for the analysis presented in this book. I outline a definition of AI and some of the key technical characteristics of this technology that are relevant for an ethical analysis of its use in defence, focusing particularly on the predictability problem, in section 2. The predictability problem opens the door to the "known unknown" risks of AI. There is growing concern that the use of unpredictable AI systems to inform high-stakes decisions may lead to severe negative consequences (Holland Michel 2020b), ranging from security risks for critical infrastructure and risks to the rights and well-being of individuals, to conflict escalation or diplomatic fallout and violation of the principles of Just War Theory. Thus, the predictability problem is central to any analysis of ethical implications of AI in defence; this why I introduce it here. I then introduce the methodology of *levels of abstraction* (Floridi 2008), in section 3. This methodology will enable us to determine the scope of the analysis proposed in the book, which I outline in section 4. Here, I describe the three categories of uses of AI in defence and offers some examples of the relevant ethical challenges they pose. These categories provide the backbone of this book, thus I develop an in-depth analyses of their ethical implications in the following chapters. I conclude the chapter in section 5.

2. Artificial Intelligence and the Predictability Problem

AI can be defined in different ways, for example by focusing on the automation of intelligent behaviour or on the design of intelligent agents and computational models of human behaviour. For the purposes of this book, we disregard the specific technical aspects of a system—for example, whether it is a statistical or a subsymbolic system—and focus instead only on the specific features of AI systems that give rise to ethical challenges. In the rest of the book, when referring to AI, I will be referring to "a growing resource of interactive, autonomous, and self-learning *agency*, which can be used to perform tasks that would otherwise require human intelligence to be executed successfully" (Floridi and Cowls 2019, emphasis added). The combination of agency, autonomy, and learning skills is the crux of the matter, as it underpins both the beneficial and the problematic uses of AI. It is also the source of the predictability problem: the limited certainty with which one can answer the question "What will an AI system do?"

Unpredictable systems are not a new issue. They are common in mathematics and physics, and limits on the ability to predict the outcomes of artificial systems have been proven formally since the 1950s (Rice 1956; Moore 1990; Musiolik and Cheok 2021). Wiener and Samuel debated the predictability of AI systems in a famous exchange in 1960 (Wiener 1960; Samuel 1960), with Wiener attributing the lack of predictability to the learning abilities of these systems, and noting "as machines learn they may develop unforeseen strategies at rates that baffle their programmer" (Wiener 1960, 1355). Developments in AI research have proven Wiener correct. Consider, for example, reward hacking, which has been reported in the literature as one of the causes of the predictability problem. In this case,

> autonomous agents optimize the reward function [given by the designers]. . . . When designing the reward, we might think of some specific training scenarios, and make sure that the reward will lead to the right behavior in those scenarios. Inevitably, agents encounter *new* scenarios (e.g., new types of terrain) where optimizing that same reward may lead to undesired behaviour. (Hadfield-Menell et al. 2020, 1)

The predictability of AI systems tends to be framed today as either a technical problem—emerging because of the technical characteristics of AI systems (International Committee of the Red Cross 2019; Boulanin et al. 2020; Defense Innovation Board 2019)—or as an operational problem resulting from the interaction of the system with its environment of deployment (International Committee of the Red Cross 2019; Docherty 2020).

The technical predictability of an AI system is assessed in terms of the degree of consistency between its past, current, and future behaviours (Holland Michel 2020a). Key aspects monitored here are data and concept shift; how often and for how long the outputs of a system are correct; and whether the system can scale up to handle correctly data that diverges from training and test data (Boulanin et al. 2020; Collopy, Sitterle, and Petrillo 2020; Defense Innovation Board 2019). Technical predictability also depends on properties such as the interpretability, transparency, explainability, and robustness of AI systems (Holland Michel 2020a; Rudin, Wang, and Coker 2020). It is worth stressing here that limited predictability in an AI system is not the same as limited robustness. Robustness refers to the ability of an AI system to produce correct outcomes even when fed incorrect data, whereas predictability refers to the probability that the system will act as expected, independent of the correctness of its outcomes. However, as the focus on technical predictability is usually centred on system error or manipulation, the distinction between unforeseen but correct outcomes and unforeseen and incorrect outcomes is often lost. In this case, while neither system is predictable, the first is at least robust.

Operational predictability refers to the degree to which the actions of a system can be anticipated once it is deployed in a specific environment. In this sense, "all autonomous systems exhibit a degree of inherent operational unpredictability, even if they do not fail or the outcomes of their individual action can be reasonably anticipated" (Holland Michel 2020b, 5). Technical and operational predictability is impacted by a large set of variables: the technical features of the system, the characteristics of the context of deployment, interactions with other systems, and how well the operator understands the way the system works and, in the defence domain, the behaviour of the opponents. These variables may change and interact together in complex ways, making it difficult to predict all possible outputs of an AI system and their effects.

The predictability problem is multidimensional. This is why the distinction between technical and operational predictability of AI system is not really tenable in practice, because technical and operational factors all contribute to determine the behaviour of an AI system. Elsewhere (Taddeo et al. 2022), I have argued that the predictability problem is best framed instead as a range of outputs that vary in their predictability, resulting from the combination and interaction of technical, security, and operational aspects. To this end, I have defined the lower and upper limits of the predictability problem:

> *Minimally*, given an ideal scenario where no errors at design and development stages can be assumed or detected, once deployed an AI system may still develop correct (and yet unwanted) outcomes, which were not foreseeable at the time of deployment.
>
> *Maximally*, given the multi-faced processes of design, development, and deployment of AI systems, the opaqueness of these systems, their adapting capabilities, and the possible complexities of the environment of deployment, it is neither possible to account for all sources of errors and manipulation of a system nor for all possible emerging behaviours—whether beneficial or not—of an AI system that these errors may prompt. (Taddeo et al. 2022, 15)

One clarification is important before we consider in more detail some of the key causes of the predictability problem: the unpredictability of an AI system is not boundless. It is limited by the system's affordances, that is, the set of hardware and software specifications that determine the range of possible actions of a machine. For example, an unsupervised system designed and developed to distinguish pictures of horses from dogs may be unpredictable with respect to the way it handles the visual inputs and the final selection of pictures. But there is no concern that the system will develop an unpredicted behaviour outside its affordances and produce a new type of outcome, like drawing a picture of a horse or a dog. It follows that as the affordances of a system increase in both degree and complexity, the wider will be the range of unpredictable behaviours that the system may show upon deployment.

2.1. Human-Machine Teaming

At the lower end of the spectrum outlined by the minimal-maximal definition above, we can imagine a system that results from a flawless design and development process and yet still shows some unforeseen outcomes, as a consequence of its learning abilities and interactions with the environment or the mode of deployment. I shall focus here on the mode of deployment because, contrary to the complexities of the environment, the environment of deployment can be decided ex ante and designed to mitigate the risks of unpredicted outcomes. At the same time, the mode of deployment of AI systems becomes a crucial aspect to consider as we move from the use of AI as a tool to the integration of AI as an artificial agent in a professional team.

In human-machine teaming that includes AI systems (HMT-AI) technical as well as cultural, ethical, legal, and cognitive factors can all contribute to unforeseen outcomes (Andras et al. 2018; Chopra and Singh 2018; Ehsan and Riedl 2020; Makarius et al. 2020; NIST 2022). HMT-AI marks a pivot from previous approaches to AI deployment, which assumed a clear division of labour between human and artificial agents, relied on low levels of automation, and assigned the processing of multiple sources of information only to humans (Shaw et al. 2010; Walliser et al. 2019; Woods, Patterson, and Roth 2002). HMT-AI focuses on producing joint intelligence systems, whereby the tasks of human experts and AI systems are distributed to create agile team processes and facilitate emergent capabilities (O'Neill et al. 2020). HMT-AI is currently deployed in several domains (Lavin et al. 2021; Scherrer et al. 2022), including logistics (Stowers et al. 2021), urban search-and-rescue teams, advanced surgical operations teams (You and Robert 2016), and cybersecurity (Stevens 2020), where researchers combine the experience and intuition of experts with ML techniques to create a system that is capable of detecting and defending against unforeseen attacks (Veeramachaneni et al. 2016). Defence operations also rely on HMT-AI (Konaev and Chahal 2021; Lopez 2022) to support decision-making, including bulk data analysis and predictive analytics in intelligence operations to enable the analysis of high dimensional data that would otherwise remain untapped (National Academies of Sciences, Engineering, and Medicine et al. 2022).

Human trust in artificial agents is a key and problematic component of HMT-AI, and one that can also worsen the predictability problem, when human agents have inappropriate levels of trust in the artificial agent. Operators may under-trust the artificial agent, resulting in disproportionate supervision of its actions or not using the AI system to its full potential. This creates opportunity costs, as the human agents waste time and resources in supervising the artificial agent rather than performing tasks the AI cannot do, but has little bearing on the predictability problem. This is not the case with over-trust, when human agents delegate too much to the artificial agent and supervise it too little, resulting in unnecessary risks. In high-stakes decision-making contexts, over-trust aggravates the risks of unpredictability and can lead to adverse outcomes, including elevated risks for the humans interacting with it. For example, a 2016 study describes an experiment involving 42 volunteers in a simulated fire emergency with a robot guide tasked with leading them to safety (Robinette et al. 2016). Nearly all the participants followed the robot blindly as it committed several fatal mistakes for which it provided neither explanation nor warning.

Over-trust can also generate a *trust and forget* dynamic (Taddeo 2017c), in which the human has the highest level of trust in the artificial agent, does not supervise its performance, and overlooks its (potentially erroneous) actions, disregarding its capabilities and limits, and accepting uncritically its outcomes. This over-trust can result from automation bias in AI—that is, the tendency of human agents to over-rely on AI outcomes (Goddard, Roudsari, and Wyatt 2012). As Struß argues, this bias and the risk of over-trust become more problematic as AI systems become more complex. This is because the human and the artificial agents in HMT-AI have different decision-making processes, and the human agent may be unable to scrutinise or understand how the artificial agent reaches its decisions, but nevertheless still rely on it due to the automation bias.[3]

Training and experience-building programmes can, however, help human agents to calibrate their expectations and form a more accurate understanding of the system's general behaviour policy, as well as overcome interface and trust issues. However, when we consider HMT-AI, these training programmes require novel concepts, methods, and standards (Laird, Ranganath, and Gershman 2019; Lavin et al. 2021; National Academies of Sciences, Engineering, and Medicine et al. 2022). Unfortunately, the HMT-AI literature still builds on structures developed for more traditional HMT with lower levels of automation and learning abilities. This presents a clear problem, given that the characteristics and dynamics of HMT do not map perfectly to those of HMT-AI—including how roles and objectives are assigned dynamically in new contexts of deployment, how shared representations are developed, and how responsibility is assigned.

---

[3] It is worth noting that every application of AI will have different data models or ML approaches (from supervised learning to reinforcement learning), or sensors, and this will affect the level of automation embedded in the AI system. In turn, the level of automation in the system will affect the ability for human agent to scrutinise and understand its outputs.

The distinct training approaches demanded by HMT-AI remain underexplored (McNeese et al. 2021; O'Neill et al. 2020). In designin new training it is crucial to include regular uncertainties and perturbations to help *both* humans and artificial agents construct well-rounded representations of each other's decision-making criteria (Niu, Paleja, and Gombolay 2021; Shih et al. 2021). For example, studies on trust in emergency guide robots (Robinette, Howard, and Wagner 2017) have highlighted the potential benefit for humans in HMT-AI of experiencing wrong behaviour from the robot prior to use in real situations, so that the human can gain some awareness of, and adapt to, the machine's imperfections. At the same time, the improved mental models developed by human agents during training sessions can be used to improve the performance of collaborative artificial agents or the interface that facilitates communication between agents, creating a feedback loop (Klamm et al. 2019). The value of this has been discussed for HMT-AI in Real Time Strategy games (A. Anderson et al. 2020), military war-gaming (Schwartz et al. 2020), autonomous flight teaming (Tossell et al. 2020), and cybersecurity (Buchanan and Imbrie 2022; Ding et al. 2019; Gomez, Mancuso, and Staheli 2019).

2.2. Machine Learning

Looking now at the higher end of the range of the predictability problem outlined above (i.e., the maximal limit), we come to consider the unforeseen outcomes resulting from combinations of errors during the design and development stage. Three sources of errors are crucial here, originating in ML models, data curation, design and development processes.

One of the main issues associated with the best-performing families of ML models (e.g., neural networks and boosted trees) is that their complexity makes it difficult to assess whether the models are generalising appropriately on data outside the training distributions. Model confidence is the most common approach in modern ML to deal with the uncertainty associated with generalisation, by assessing the different uncertainties that characterise the model and its operational environment (Hüllermeier and Waegeman 2021). However, model confidence is often not robust from a statistical point of view. Deep neural networks, for example, have been proven to be overconfident, possibly leading to high-confidence mistakes or accidently concealing adversarial attacks being conducted on the model (ENISA 2020a). Confidence levels therefore have to be adjusted to outputs, which will complicate (and may perturb) subsequent processes.

At the same time, even for the best-performing AI models, training outcomes are not necessarily indicative of the capabilities of a system in the real world, where deployment conditions will diverge from training ones, and new data falls outside the distributions of the training dataset. Cases of deep neural network models failing to generalise appropriately outside training conditions are often reported in the literature (Nguyen, Yosinski, and Clune 2015; Athalye et al. 2018. For example, in computer vision, it is difficult to analyse images where there is a noisy context or contextual confusion of extraneous pixels or light. AI systems have been shown to be susceptible to minor changes, down to pixel level, with minor variations leading a system to misidentify stripes as school buses (Nguyen, Yosinski, and Clune 2015). These limitations have been shown to be exploitable in AI-enabled multi-domain operations in the military (Jia et al. 2022; Savas et al. 2020).

2.3. Data Curation

Data curation is a crucial step in the development of AI systems, with data labelling a key aspect when considering the predictability problem. Labels attach meaning to training data, enabling a machine to learn from it. There are different methodologies available for labelling, all of which have important limitations that could lead to unpredictable system outcomes. For example, data labellers may reproduce bias (Bekele, Narber, and Lawson 2017; Bekele et al. 2018), creating skewed training datasets that will impact the performance of the AI system and could lead to unpredicted outcomes. Other forms of labelling, like consensus voting labelling, may improve overall quality but at a higher cost than for other forms of labelling. In some cases, it may be possible to create synthetic labelled data. This requires vast processing power, however, and comes with elevated error potential (IBM 2021).

Data cleaning is another form of curation. It removes duplicates, uninformative features, and outliers from a dataset (Tobin 2022) with the aim of removing noise and improving the model's performance. However, data cleaning introduces the risk of removing meaningful data points. This could lead to unintended outcomes in the application phase, if the resulting clean dataset has been stripped of important information that is useful for testing model behaviour. When datasets are constructed from multiple sources of input, some of which may be incompatible, the data-commingling problem emerges. This is the case, for example, when different sensors are used without having been calibrated or normalised to produce the same values, leading to inconsistent, incomplete, and inaccurate datasets and to unreliable outcomes.

Data shift is the extent to which system outcomes have moved off course as a result of external factors, leading to a change in data distribution (Sarantitis 2020). Building an AI model requires identifying predictable relationships between input and target variables, with an expectation that the same data distribution will elicit similar results. However, unpredictable factors in the real world can change inputs, dataset quality, data capture (e.g., polling frequency), or even the underlying patterns forming relations between input and output data (Sarantitis 2020).

As well as the possibility of introducing unwanted errors in the outputs of an AI system, data curation poses two important operational challenges vis-à-vis the predictability of a system's behaviour. The first is the operational pay-off (effort vs. resulting efficiency) of performing data curation, and the second emerges if there is a lack of standards and automated mechanisms to improve data quality. Some of the key dimensions of data quality are completeness, accuracy, uniqueness, timeliness, consistency, and validity. However, these dimensions and their relative importance may vary depending on the different contexts of use and the related purpose. The definition of quality standards and governance mechanisms for unstructured data are still lacking. This is problematic, as these data are generally used in AI models. As the EU Agency for Fundamental Rights has highlited (FRA 2019), this lack of quality specifications and guidance leads to lack of efficient standards, tools, and mechanisms[4] to evaluate data robustly and check whether they are fit for purpose. These limits to assessment of the quality of training data can result in noisy, error-ridden, and inconsistent datasets, which may lead to unpredicted behaviour at a system level. If unchecked, data-related errors and uncertainties continue to accrue and propagate across the various elements of an AI system.

2.4. Technical Debt

In software development, 'technical debt' is a metaphor used to refer to long-term software issues and costs stemming from forgoing best practices at the development stage in favour of easier and quicker solutions. Best practices commonly implemented in modern software development—like version control and unit and

---

[4] "A Data-Centric View of Technical Debt in AI", Data-Centric AI website, https://datacentricai.org/data-in-deployment/ (accessed August 3, 2024).

system testing—are not so easily translated to the AI domain, due to the lack of standard procedures and frameworks and the inherent difficulty of defining robust tests for AI models (Sculley et al. 2015). Mitchell et al. (2019), among others, have proposed solutions such as frameworks for packaging and shipping production models using model cards describing quantitatively their design space, key metrics, and known limitations, but these have yet to see widespread adoption.

The lack of commonly accepted versioning and testing tools in the AI domain generally translates into a patchy adoption of Continuous Integration/Continuous Delivery (CI/CD) practices. Reliable CI/CD pipelines require, for example, extensive versioning. Not being able to version an AI system reliably can cause robustness issues, this why it is a contributing cause of the predictability problem in its maximal definition. At the same time, the numerous and interrelated components of an AI system make its abstraction boundaries hard to control. These aspects become more problematic as ML models continue to evolve after deployment. Insofar as it hinders the reliability and traceability of AI systems, technical debt inhibits the ability of observers to predict a system's outcomes.

Having outlined the key aspects of the predictability problem here, I will focus on its ethical implications for the use of AI in defence in the rest of this book, particularly in Chapters 4–8. The next two sections of this chapter outline the methodology and the scope of the analysis.

3. The Methodology of Levels of Abstraction

The analysis proposed in this book uses the methodology of levels of abstraction, or LoAs (Floridi 2008). LoAs are used in systems engineering and computer science to design models of a given system (Hoare 1972; D. Heath, Allum, and Dunckley 1994; Diller

1994; Jacky 1997; Boca 2014). They are also widely used in digital ethics (Floridi 2008; Floridi and Taddeo 2016) and have been applied to address several key issues, including identifying the responsibilities of online service providers (Taddeo and Floridi 2015), offering guidance on the deployment of tracing and tracking technologies during the Covid-19 pandemic (Morley, Cowls, et al. 2020), analysing the possibilities of deterrence in cyberspace (Taddeo 2018c), and considering the ethical implications of trust in digital technologies (Taddeo 2017b).

The method rests on the assumption that any system can be observed by focusing on certain specific properties while disregarding others. The choice of these properties, that is, the observables, depends on the aim of the observer. For example, for an engineer interested in maximising the aerodynamics of a car, the observables may be the shape of its parts, their weight, and the materials. For a customer interested in the aesthetics of the same car, the observables may be instead its colour, the car's interior, and the overall look. The engineer and the customer observe the same car (system) at different LoAs, which will enable them to define different models of that car.

Thus, a LoA is defined as a finite but non-empty set of observables accompanied by a statement of what precise feature of the system the LoA stands for. A LoA does not reduce a car to the aerodynamics of its parts or to its overall look. Rather, a LoA is a tool that helps to make explicit the perspective of the observer of the system and helps constrain it to only those elements that are functional for that particular observation and aim. A LoA can have a lower or higher granularity, in which "The quantity of information in a model varies with the LoA: a lower LoA, of greater resolution or finer granularity, produces a model that contains more information than a model produced at a higher, or more abstract, LoA" (Floridi 2008, 315). When considering the ethical challenges of AI used in defence, one can focus on different LoAs. For example, one may decide to consider only ethical problems emerging during the design stage and disregard the development and deployment stages of the AI life cycle. Similarly, ethical analyses may focus only on the intention of use or on the effects of use of AI in this domain. The choice of the LoA is purpose oriented: there is no correct or incorrect LoA *per se*, but only correct or incorrect LoAs given the goal of the observer.

Given the goal of this book, I will adopt two LoAs: $LoA_{purpose}$ and LoA_{ethics}. The observables of $LoA_{purpose}$ are the immediate purposes of use of the AI. The observables of LoA_{ethics} are, for any given immediate purpose of use, the aspects of the design, development, and deployment of AI that may lead to un/ethical consequences. It is worth stressing that the purpose of use is not the *function* of an AI system, given that a system with the same function may pose different ethical problems when deployed for different purposes. For example, an image recognition system poses different problems when used for monitoring marine life than when implemtented on a loitering drone. The choice to focus here on purposes of use rather than on the function of the technology rests on two reasons: the malleability of digital technologies and the overall goal of this book.

Malleability refers to the fact that digital technologies, even the most sophisticated, can easily be repurposed (Moor 1985, 269). Because of their malleability, the ethical challenges of digital technologies (AI in particular) are not defined by their design function as much as by the purpose for which these technologies are deployed. Within the defence domain, these purposes can be identified clearly and are likely to shape both current and future uses of AI. This is why the focus on purpose is more appropriate for developing an ethics of IA in defence.

With respect to the goal of this book, this is not to define a comprehensive taxonomy of AI technologies and their related ethical implications, but to offer criteria to identify the ethical problems linked to use of AI in the defence domain, analyse their

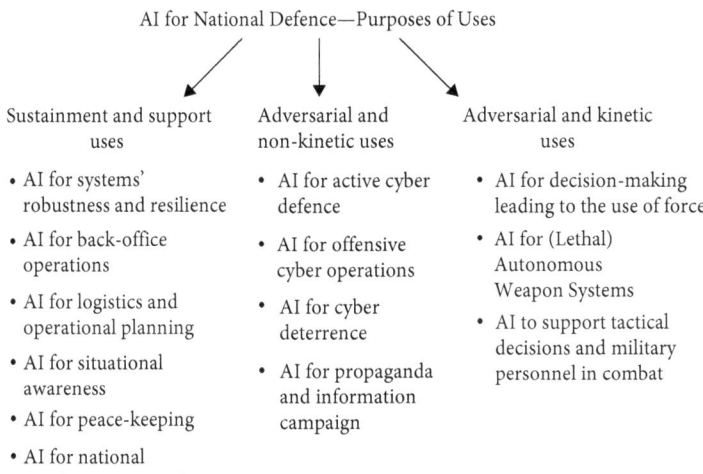

Figure 1.1 The three purposes of use of AI for national defence examined in this book. This figure is adapted from Taddeo, McNeish, et al. 2021, 1710.

implications, and provide ethical solutions and effective guidance to address them. This is why the focus on purpose rather than function is appropriate here.

Using these LoAs, we can now define the scope of the ethical analysis of AI in defence. Three purposes of AI in defence will be discussed in this book: sustainment and support; adversarial and non-kinetic; and adversarial and kinetic (Figure 1.1).

4. Ethical Problems of Using AI for Defence Purposes

The three purposes of use of AI in the defence domain set out above present increasingly difficult ethical problems as we move from sustainment and support uses to adversarial and kinetic uses (see

Figure 1.1). This is because alongside the ethical problems related to the use of AI (e.g., transparency and fairness), we also need to consider the ethical problems related to adversarial (kinetic and non-kinetic) uses of this technology and their disruptive and destructive impact.

As shown in Figure 1.2, each category of use presents its own specific ethical risks, but also inherits the ones from the categories to its left. For example, adversarial and non-kinetic uses of AI pose risks to individual rights as well as of escalation. Adversarial and kinetic uses of AI pose risk for transparency and human autonomy, which appear in the sustainment and support category, alongside the protection of rights, while also respecting the principles of Just War Theory, military virtue, human dignity, and stability. The following sections will consider in more detail some of the key ethical risks of each purpose of use.

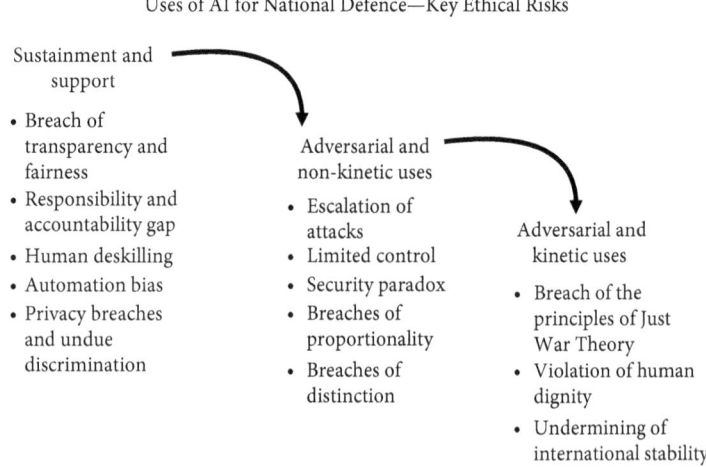

Figure 1.2 A map of the ethical risks related to each category of use of AI in defence.

4.1. Sustainment and Support Uses of AI

The use of AI systems for non-adversarial defence purposes varies from applications in cybersecurity (US Army 2017), where AI plays an ever-growing role to ensure systems robustness and resilience, to AI-based drones capturing video reconnaissance, radio-frequency identification (RFID) tags on food supply (Lysaght, Harris, and Kelly 1988; Fraga-Lamas et al. 2016; Schubert et al. 2018), and use of AI for intelligence analysis (Blanchard and Taddeo 2023).

For nations with adequate capabilities, AI systems are likely to reach full integration into national defence capabilities to support back-office, logistics, and intelligence tasks, as well as to improve the robustness of the digital infrastructure. Even when this use of AI poses no risks to individual rights and is not related to the use of force, it still poses important ethical challenges. Consider, for example, the use of AI to enhance system robustness. AI can help with verification and validation of software, liberating human experts from tedious jobs, and offering faster and more accurate testing of systems (King et al. 2019). For example, generative models like large language models (LLMs) have shown great potential for vulnerability scanning.[5] In this sense, AI can take software testing to a new level, making systems more robust. However, we should be careful about the way we use AI in this context, for delegating testing to AI could lead to a complete deskilling of defence personnel deployed for verification and validation of systems and networks, eventually resulting in a lack of control of this technology.

AI is also increasingly deployed for threat and anomaly detection, in which existing security data are used to train a system in pattern recognition. For example, in April 2017 the software firm DarkTrace launched Antigena, which uses ML to spot abnormal

---

[5] "Security Implications of ChatGPT", Cloud Security Alliance, published February 8, 2023, https://cloudsecurityalliance.org/artifacts/security-implications-of-chatgpt/ (accessed August 2, 2024).

behaviour on an IT network, shut off communications to that part of the system, and issue an alert. These services analyse malware and viruses, and some can quarantine threats and portions of the system for further investigation. In certain cases, threat scanners have access to files, emails, mobile and endpoint devices, or even traffic data on a network. AI can also be used to authenticate users by monitoring behaviour and generating biometric profiles based, for example, on the unique way a user moves her mouse around ("BehavioSec: Continuous Authentication Through Behavioral Biometrics" 2019). Sometimes this may involve tracking "sensor data and human-device interaction from your app/website. Every touch event, device motion, or mouse gesture is collected".[6]

The risk is clear. AI can improve system resilience to attacks, but this requires extensive monitoring of the system and of its users and comprehensive data collection to train the AI. This devalues user privacy, exposes users to additional risks should data confidentiality be breached, and creates a mass-surveillance effect (Taddeo 2013, 2014b). Indeed, the sustainment and support uses of AI in defence and security pose ethical challenges similar to those reported for other domains, for example concerning privacy breaches. However, it is important to note that these problems need to be addressed with specific regard to the defence domain, which may require a complex balancing of state interest, use of force, national defence, and respect for individual rights.

AI is also deployed to enhance situational awareness and intelligence analysis (of which more in Chapter 3). Timely, situational awareness is important for enhancing preparedness and pre-empting threats. However, enhancing such awareness using AI can be ethically problematic, especially given the hybrid nature of threats and different variables at play. This is because the threats, which may be hybrid in nature, may also coincide with changing facets in the political, economic, strategic, cultural, and

---

[6] "Denuvo Unbotify", ir.deto, http://www.unbotify.com (accessed June 5, 2024).

social circumstances operating around the defender, and attacks can be initiated by actors working with changing allies, interests, resources, and methods. This requires always-on, real-time analytics and anomaly detection capabilities. AI offers much to this end, as it enables the analysis of great volumes of data, but the key challenge is to ensure that large-scale data collection and analyses are kept in balance with key regulations and ethical values, to avoid undermining civilian trust in defence and security institutions, for example through excessive, undue surveillance or discriminatory systems.

The analysis of large volumes of data also enables AI to extract information to support logistics and decision-making, as well as foresight analyses, internal governance, and policy. These uses of AI facilitate timely and effective management of both human and physical resources, improve risk assessment, and support decision-making. For example, it has been reported that military officers could have only between eight and 10 minutes to decide whether a missile launch represents a threat, share the findings with allies, and decide how to respond.[7] AI would be of great help in this scenario, for it could integrate real-time data from satellites and sensors and process key information that may contribute to the decision-making process.

The good potential of AI to aid situational awareness is coupled to serious ethical risks, including undue discrimination if AI systems are biased, lack of transparency and accountability, automation bias, and responsibility gaps, posing questions concerning the way these systems are designed and integrated into the decision-making process.

---

[7] "The Next Major Defense Challenge", KPMG, June 11, 2018, https://kpmg.com/ph/en/home/insights/2018/06/the-next-major-defense-challenge.html (accessed August 3, 2024).

4.2. Adversarial and Non-kinetic Uses of AI

As threats against the state escalate, so does the need for defence strategies to meet them. This includes cyber threats. The UK and the US have employed active cyber defence (or "defend forward") approaches to enable computer experts to neutralise or distract viruses with decoy targets, and to break back into a hacker's system to delete data or to destroy it completely. In 2016, the UK announced £1.9 billion of investment and a five-year plan to combat cyber threats, with the commitment being increased to £2.6 billion in 2022.[8] In 2020, the UK also established the National Cyber Force as a joint initiative between the Ministry of Defence and Government Communications Headquarters, tasked with targeting hostile foreign actors. On an international scale, NATO can now rely on sovereign cyber effects (i.e., cyber-attacks launched by its member states) in response to cyber-attacks, as agreed at the Brussels Summit,[9] to enable the alliance to punish (attributed) cyber-attacks and deter attackers from striking again.

AI will continue to revolutionise these activities. Cyber-attacks and responses will become faster, more precise, and more disruptive. AI will expand the targeting ability of attackers, enabling them to use more complex and richer data to choose targets and design attacks, and AI within malware can change the nature and delivery of an attack. AI-enabled cyber weapons have been prototyped already, including autonomous malware intended to corrupt medical imagery and attack autonomous vehicles (Mirsky et al. 2019; Zhuge et al. 2007). Similarly, IBM has created a prototype autonomous malware, DeepLocker, that uses a neural network to select its targets

---

[8] "Government Cyber Security Strategy", UK Government website, https://assets.publishing.service.gov.uk/government/uploads/system/uploads/attachment_data/file/1049825/government-cyber-security-strategy.pdf (accessed August 3, 2024).

[9] "NATO's Rolemin Cyberspace", published on 12 February 2024, https://www.nato.int/docu/review/articles/2019/02/12/natos-role-in-cyberspace/index.html (accessed August 3, 2024).

and disguise itself until it reaches its destination ("DeepLocker" 2018). Autonomous and semi-autonomous cybersecurity systems endowed with a "playbook" of predetermined responses to an activity, constraining the agent to known actions, have been available on the market for a few years now ("DarkLight Offers First of Its Kind Artificial Intelligence to Enhance Cybersecurity Defenses" 2017). Autonomous systems that are able to learn adversarial behaviour and generate decoys and honeypots, thus luring threat actors ("Acalvio Autonomous Deception" 2019), are also being commercialised. Research has also shown that large language models (LLMs) can exploit autonomously real-world, zero-day vulberabilities, with serious cybersecurity consequences (Fang et al. 2024).

As state actors acquire and develop AI capabilities for national defence, so do their opponents (Taddeo and Floridi 2018a). As, we shall see in Chapter 4, this can lead to a security paradox, whereby state actors accruing AI capabilities may be perceived as a threat by others; this may lead to international skirmishes, which may cause an intensification of cyber-attacks and responses, positing risks of escalation leading to kinetic consequences (Taddeo 2018c). To mitigate these risks, it is vital that uses of AI respect key principles of Just War Theory, which underpins international regulations, such as the United Nations Charter,[10] The Hague and Geneva Conventions,[11] and International Humanitarian Law,[12] and sets the parameters for both ethical and political debates on waging cyber conflicts. It is crucial that the deployment of AI for adversarial and non-kinetic purposes respects the principles of proportionality of responses, discriminates between legitimate and

---

[10] UN Charter, UN website, https://www.un.org/en/sections/un-charter/un-charter-full-text/ (accessed August 3, 2024).

[11] Collection Military Legal Resources, Libraray of Congress website, https://www.loc.gov/rr/frd/Military_Law/pdf/ASubjScd-27-1_1975.pdf (accessed August 3, 2024).

[12] International Committee of the Red Cross website, https://www.icrc.org/en/doc/resources/documents/misc/57jm93.htm (accessed August 3, 2024).

illegitimate targets, ensures some form of redressing when mistakes are made (Taddeo 2012a, 2012b; 2014a), and maintains responsibility and control within the chain of command. Ultimately, ethical considerations on the adversarial and non-kinetic use of AI should contribute to an understanding of how to apply Just War Theory in cyberspace, and should be used to shape the debate on the regulation of state behaviour in cyberspace (Taddeo 2016a, 2017a). These points will be discussed further in Chapters 4 and 5.

4.3. Adversarial and Kinetic Uses of AI

When considering the ethical implications of adversarial and kinetic uses of AI for defence and security, the main focus of analysis has tended to be on the combination of AI with machinery that can cause lethal harm to humans and destruction to physical objects in a completely autonomous way. However, the use of AI for adversarial and kinetic purposes varies greatly, ranging from automation of various functions of a weapon system, to systems that follow the pre-programmed instructions of a human, to fully autonomous weapons systems (AWS) that can identify, select, and engage targets without direct human input. Consider for example, a system developed for the UK Royal Navy called STARTLE,[13] which supports human decision-making with situational awareness software that monitors and assesses potential threats using a combination of AI techniques. Similarly, the Advanced Targeting & Lethality Automated System[14] developed for the US Army supports human operators in identifying threats and prioritising potential targets. Reportedly, Israel has developed and deployed The Gospel

---

[13] "Complex information fusion and advanced threat warning system", Roke website, https://www.roke.co.uk/products/startle (accessed August 3, 2024).
[14] "ATLAS: Killer Robot? No. Virtual Crewman? Yes.", Breaking Defense website, Published on 4 March 2019, https://breakingdefense.com/2019/03/atlas-killer-robot-no-virtual-crewman-yes/ (accessed March 4, 2019).

and Levander, two AI systems for human target identification.[15] These uses of AI result in ethical questions and problems whose severity vary with the degree of autonomy of the AI system, its learning features, and the level of human control.

One of the key challenges in this domain is to ensure that adversarial and kinetic uses of AI respect the principles of Just War Theory, for example necessity, proportionality, and distinction. This concerns both the modes of deployment and the capabilities of the deployed systems. Procedurally, AI systems must be deployed following procedures that ensure appropriate human oversight and ability to override AI decisions. Technically, for example, AI systems must be able to distinguish between combatants and noncombatants and must recognise the generally accepted signs of surrender that operate in armed conflict. Some have argued that this is problematic, given that AI is not yet able to analyse context to this degree, and in some situations its capacity to recognise legitimate targets is significantly worse than that of humans (Sharkey 2010, 2012a, 2012b; Tamburrini 2016). I agree. However, I have argued elsewhere that the state of development of AI is essentially moot; the principles of Just War Theory will always be at risk because of the inherent predictability problem of AI, and the resulting limited control of lethal autonomous weapon systems (LAWS) (Blanchard and Taddeo 2022a, 2022b). Even if AI technologies were advanced enough to interpret contexts correctly, the limited predictability of its outcomes will introduce a (non-zero) source of risk that may be considered unacceptable when contemplating kinetic uses. This is why the debate on the permissibility of LAWS needs to focus on acceptable risk thresholds more than on the state of development of AI technologies. I shall return to this in Chapter 8.

---

[15] "'Lavender': The AI machine directing Israel's bombing spree" in Gaza +972 Magazine, published on 3 April 2024, https://www.972mag.com/lavender-ai-israeli-army-gaza/ (accessed August 3, 2023)

The responsibility gap is another key ethical challenge. While it is problematic in all three categories of use of AI, it is particularly worrying when considering the adversarial and kinetic case, given the high stakes involved (Sparrow 2007b). The responsibility gap becomes even more pressing as an issue when considering the respect for opponents and their dignity. Treating opponents with respect in warfare is an important way to maintain warfare's morality (Nagel 1972), and the interpersonal relation with the opponent is key to this end. Insofar as the use of LAWS severs this relation, the question emerges as to whether the use of these systems undermines the dignity of those whom they target (and possibly also those who use them) and thus leads to a form of morally problematic killing (Asaro 2012; Docherty 2014; A. Sharkey 2019; Johnson and Axinn 2013; Sparrow 2016; O'Connell 2014; Ekelhof 2019).

Questions also arise with respect to the impact of LAWS on international stability. On the one hand, LAWS may reduce the time span of interstate hostilities and thus foster stability. They could also be an effective deterrent against possible opponents. On the other hand, LAWS may lead to unjust war and promote international instability, for example if asymmetric use of LAWS leads to the weaker side resorting to insurgency and terrorist tactics (Sharkey 2012a, 2012b). Because terrorism is generally considered to be a form of unjust warfare (or, worse, an act of indiscriminate murder), deploying LAWS may lead to a greater incidence of immoral violence. I shall return to these points in Chapters 6, 7, and 8.

5. Conclusion

Having outlined the scope and methodology of the analysis in this book, here I turn to a discussion of its goal, which is to provide a systemic ethical analysis of the uses of AI in defence. I hope the book will thereby identify opportunities and risks in this domain and offer guidance on leveraging the former and mitigating the latter. I should be honest with the reader at this point and declare that there is another goal to which the analysis presented in this book is instrumental, namely fostering a more proactive approach of the defence establishment toward the ethics of AI in defence.

Defence, whether kinetic or not, has become digital. Data and technologies to collect, analyse, and communicate—and the AI systems that increasingly support and deliver data—are key capabilities that are leading to profound changes in military doctrines and strategies. The digital revolution is transforming defence no less that it has transformed our daily lives and societies. These changes are disruptive, and if not understood properly they can lead to unwanted consequences and breaches of fundamental values and rights.

The case of cyber conflicts is emblematic here. The normative gap for state behaviour in cyberspace has developed as the result of a number of reasons, a key one being the perception of ethics as—at best—an add-on to the discussion of the digital transformation of warfare, and at worst an annoying hindrance to the need to gain advantage over the opponent. A decade on from the 2014 NATO Summit, this approach has favoured aggressive behaviour in cyberspace, where state-run and state-sponsored cyber-attacks continue to grow, with attacks like WannaCry, NotPetya, and the attack on the US Office of Personnel Management[16] causing extensive damage to civilian infrastructures and infringing fundamental rights like privacy and safety and aggravating geopolitical tensions (Taddeo 2017a).

As AI empowers defence forces, the need to regulate its uses becomes even more pressing. Indeed a number of defence institutions have put an increasing focus on the ethics of AI. However, to be effective, these efforts need to be grounded in an

---

[16] https://www.reuters.com/world/us/data-237000-us-government-employees-breac hed-2023-05-12/#:~:text=Two%20breaches%20at%20the%20U.S.,5.6%20million%20 of%20those%20individuals.

appropriate understanding of the conceptual changes of the digital revolution and their ethical implications. This is a well-known issue in digital ethics. As Moor put it, "Although a problem . . . may seem clear initially, a little reflection reveals a *conceptual muddle.* What is needed in such cases is an analysis which provides a coherent *conceptual framework within which to formulate a policy for action*" (1985, p. 266, emphasis added). This book will show that ethical analyses of AI in defence are *not* a mere add-on or a hindrance, but indeed are required to leverage the potential of AI in defence and to define its boundaries so as to ensure that it does not breach the very values and rights that defence institutions are created to protect.

2
Ethical Principles for AI in Defence

1. Introduction

Having outlined the methodology and the scope of the analysis proposed in this book, I shall now look at existing approaches to the ethics of AI in defence. As the impact of AI across the entire range of functions of national defence has become clear, questions about how to foster ethical use of AI technologies in this domain have become ever more pressing, leading a number of defence institutions to consider ethical risks and to define measures to mitigate them. Efforts in this area are still nascent, and up to 2023 only three defence institutions—the US Department of Defence (DoD) (Defence Innovation Board 2020b), UK Ministry of Defence (MoD) (Ministry of Defence 2022), and NATO[1]—have adopted official ethical principles for AI (see Table 2.1). The three sets of principles show a remarkable consistency, in terms of both content and limitations. Two of these limitations are of particular relevance for this chapter.

The first limitation comes with the approach underpinning these principles: responsible AI (RAI), which aims to foster responsible design, development, and deployment of AI systems; that is, a responsible life cycle. I will return to this point in section 3. I will simply say here that this approach is not best suited for the use of AI in high-risk domains. It builds on the responsible research and innovation approach (RRI; see, e.g., Stilgoe, Owen,

---

[1] https://www.nato.int/docu/review/articles/2021/10/25/an-artificial-intelligence-strategy-for-nato/index.html.

The Ethics of Artificial Intelligence in Defence. Mariarosaria Taddeo, Oxford University Press.
© Oxford University Press 2024. DOI: 10.1093/oso/9780197745441.003.0002

Table 2.1 A summary of the principles to inform the ethical use of AI in defence as provided by the US DoD, the UK MoD, and NATO

Principles	US DoD	UK MoD	NATO
	Responsible	Human-centred	Lawfulness
	Equitable	Responsibility	Responsibility and accountability
	Traceable	Understanding	Explainability and traceability
	Reliable	Bias and harm-mitigation	Governability
	Governable	Reliability	Bias mitigation

and Macnaghten 2013), which aims to foster critical reflection on the potential implications of research, rather than offering concrete guidance as to what one ought, or ought not, to do to mitigate ethical risks in specific cases. However, concrete guidance is *precisely* what is needed to promote ethical use of AI in defence.

The second limitation concerns the prescriptions of the principles, which at times sound obvious to the point of being redundant. For example, one of the NATO principles states that the use of AI should be "legal", the US principles stress that AI should be used in such a way to be "governable", the UK principles state that the use of AI should be "human-centric". There is nothing controversial about the ideas that AI systems should be legal, governable, and human-centric—indeed one might ask what the alternative would look like. Here, the choice to focus on obvious points come at the cost of providing more insightful AI ethics principles focusing clearly on the risks that AI poses to values and rights of democratic societies and on domain-dependent principles, particularly those articulated in Just War Theory. This chapter has the goal of proving a set of alternative principles designed with this approach in mind and which overcomes the limitations of existing ethical principles for AI in defence and of offering a methodology to translate these principles into practices.

The second limitation of the principles is also a consequence of the very high level at which they are stated. Some have argued that such a high level is a crucial shortcoming, making these principles too abstract to guide the actual design, development, and use of AI systems in practice (Coldicutt and Miller 2019; Peters 2019) or to inform decision-making. I do not myself interpret the high level of AI ethics principles as a drawback. These are foundational principles, not guidelines; they offer a domain-dependent framing of the values at stake. In doing so, they are essential to orientate efforts to ensure ethically sound uses of AI, but alone are not sufficient to this end. That is, these ethical principles provide a compass, not a map. Criticisms questioning the efficacy of these principles because of their high level are mistaken if they understand such principles to be necessary and sufficient to inform an ethically sound AI life cycle. Rather, ethical principles are a part of more systemic efforts needed to foster an ethically sound AI life cycle in high-risk domains.

To be effective, AI ethics principles need to be coupled with appropriate methodologies to offer domain-specific guidance as to how to interpret and apply them (Taddeo and Floridi 2018b), in order to ensure that every step of the AI life cycle respects the ethical principles of the organisation overseeing and developing it. Recognising this need, the AI ethics literature has shifted from the *what* to the *how* (Floridi 2019, 185), resulting in a growing body of work that focuses on developing AI ethics tools and processes to implement AI ethics principles.[2]

However, by focusing directly on tools and applicable solutions, the relevant literature leaves unaddressed crucial normative questions concerning the implementation of ethical principles. For example, when applied to specific cases, AI ethics principles

---

[2] Morley et al. (2020) have compiled a taxonomy of these tools.

may generate tensions requiring trade-offs that cannot be resolved through recourse to the principles or tools alone (Whittlestone et al. 2019). Consider the case where transparency may conflict with national security issues. In a similar way, the choice of an AI ethics tool will involve normative decisions, which cannot be made by referring only to the principles (Blanchard, Thomas, and Taddeo 2023). For example, the choice of an ethical auditing tool requires a decision as to the type of metrics, whether qualitative or quantitative, to be used for the assessment. In turn, the type of metric determines the scope of the audit. This means there is a need for guidance as to the metrics one should choose in specific situations. In short, there is a midway step between (high-level) AI ethics principles and AI ethics tools that concerns the specification of a methodology for the implementation, that is the interpretation of these principles into practices. Providing this methodology is the second goal of this chapter.

In the rest of this chapter, I will review the AI ethics principles adopted by the US defence industry, in section 2, and the proposed methodology to implement them, in section 3, in order to extract valuable lessons before offering a new set of principles for AI ethics in defence. I outline these principles in section 4. I will introduce a methodology to extract ethically sound, feasible, and effective guidelines for the use of AI in defence from the proposed AI ethics principles, in section 5. I conclude the chapter in section 6.

2. Ethical Principles for the Use of AI

Of the ethical principles released officially by a defence institution, the ones published by the US DoD (Defense Innovation Board 2019) are the only ones accompanied by a document (Defense Innovation Board 2019) outlining the rationale underpinning them and offering recommendations for governance solutions to implement the principles (henceforth: supporting document). The

US DoD is also the only defence institution that has published official guidelines for the operationalisation of the principles (see section 3).[3] For this reason, I will focus on the US DoD's principles alone and disregard the principles issued by the UK MoD and NATO, which are similar for scope and content but less detailed. US DoD offers five AI-level principles, which prescribe that the use of AI in defence should be responsible, equitable, traceable, reliable, and governable. The following subsections analyse each principle in turn to consider valuable aspects and limitations.

2.1. Responsible Uses of AI

The first principle in the Defense Innovation Board document mandates responsible uses of AI. It states:

> human beings should exercise appropriate levels of judgment and remain responsible for the development, deployment, use, and outcomes of DoD AI systems. (Defense Innovation Board 2019, 8)

This principle is uncontroversial and coherent with the principles provided by a number of ethical frameworks (Department for Digital, Culture, Media & Sport 2018, 5; Gavaghan et al. 2019, 41; Japanese Society for Artificial Intelligence 2017, 3; Ministry of Defence 2022). In the supporting document, the recommendation on how to implement this principle proposes a three-level system of responsibilities, with the first level addressing humans who control

---

[3] Both the UK's MoD and NATO have announced that they are working on developing guidelines or methodologies to interpret and apply AI ethics principles to specific cases, but have yet to publish any relevant documents at the time of writing in November 2023.

the design, requirements definition, development, acquisition, testing, evaluation, and training for any DoD system, including AI ones. (Defense Innovation Board 2019, 27)

The second level of responsibility concerns the use of AI in the conduct of hostilities (whether kinetic or not), with responsibilities assigned to the command and control structure, insofar as commanders and operators have "appropriate information on a system's behavior, relevant training, and intelligence and situational awareness" (p. 28). The third level of responsibility concerns redressing mechanisms after hostilities have ended. This level addresses both the DoD and the private sector procuring AI technology for defence. The supporting document specifies that human responsibility in this case rests on "human appropriate judgement".

This approach is correct only in part. The definition of "appropriate" judgement is vague and, therefore, problematic, especially when considering the problems posed by the lack of transparency and predictability of some AI systems. At the same time, the attribution of responsibility according to the three-level system risks dumping responsibilities on the first level (development and training), insofar as unintended consequences of AI systems may be linked back to design and development issues. This may have a detrimental effect on the way actors involved in command and control may perceive their responsibilities with respect to the use of AI.

2.2. Equitable Uses of AI

The principle of equitable uses of AI prescribes that

the DoD should take deliberate steps to avoid unintended bias in the development and deployment of combat or non-combat AI systems that would inadvertently cause harm to persons. (p. 8)

The principle focuses on issues related to fairness and justice, while avoiding the terms directly. In the supporting document, the reason for not using the term 'fairness' is that

> this principle stems from the DoD mantra that fights should not be fair, as DoD aims to create the conditions to maintain an unfair advantage over any potential adversaries, thereby increasing the likelihood of deterring conflict from the outset. (Defense Innovation Board 2019, 31)

The document goes on to say that the

> DoD should have AI systems that are appropriately biased to target certain adversarial combatants more successfully and minimize any pernicious impact on civilians, non-combatants, or other individuals who should not be targeted. (Defense Innovation Board 2019, 33)

This departure from the usual understanding of the term 'fairness' in AI systems, then, is motivated by the perceived unique nature of defence. This is misleading, as it suggests that the need to seek advantage over the opponent may justify unfair, or indeed unjust, uses of AI. But this is not the case, as a principle of fairness does not prescribe that there be zero-sum outcomes between the interests of opposite parts. Rather, it refers to outcomes that account for individual rights, duties, in a specific scenario. In this sense it refers to the principle of justice, according to which individuals should be treated the same unless they differ in ways that are relevant to the situation in which they are involved (Rawls 2005).

The principle of justice applies also in warfare. Indeed we distinguish between just and unjust conduct in defence and punish the latter. There are differences between the ways in which the principle of justice is applied in civilian and non-belligerent contexts and in hostile activities. In defence, the principle of justice needs to be balanced with that of military necessity, which allows for the use of (lethal) force (within the limits defined by the principles of proportionality and distinction) if it is deemed necessary for a legitimate military purpose, such as the defeat of the enemy or the protection of one's own forces.

Insofar as the US DoD principle focuses only on equitable uses of AI and characterises them only with respect to risks of bias and undue discrimination against DoD personnel, it misses the opportunity to consider the wider spectrum of the risks of unjust uses and outcomes of AI in defence, and to foster mitigating measures. Consider, for example, the use of LAWS, as we will see in Chapter 8, they can be used to acquire an advantage over the opponent, but given the predictability problem, there is limited control on their effects. This implies that LAWS may breach the principle of distinction and harm non-combatants. Under Just War Theory and international humanitarian law (IHL) this would be an unjust and thus impermissible use of AI; but it is not captured by the DoD principle focusing on equitable uses of AI. Attempts to inform ethical uses of AI in defence must define principles for just uses of AI that are relevant within this domain and coherent with the principles provided by Just War Theory (Taddeo 2014a). Altenatively, they risk becoming irrelevant.

2.3. Traceability

The DoD principle of traceability addresses indirectly the ethical problems posed by the lack of transparency of AI. It states that

> DoD's AI engineering discipline should be sufficiently advanced such that technical experts possess an appropriate understanding of the technology, development processes, and operational methods of its AI systems, including transparent and auditable

methodologies, data sources, and design procedure and documentation. (Defense Innovation Board 2019, 8)

Notably, the focus of the principle is not on the transparency of the technology *per se* but on defining a minimum skill level of the DoD's personnel such that they have an appropriate understanding of its AI systems, including traceability of the processes and decisions of AI systems both at development and operational stages. As specified in the supporting document, traceability at development stage refers to the collection and sharing with appropriate stakeholders of "design methodology, relevant design documents, and data sources" (p. 34), whereas at operational stage, traceability includes forms of monitoring, auditing, and transparency of processes. As specified in the supporting document:

> some systems may require not just reviews of user access, but also records of use and for what purpose. This requirement can mitigate harms related to off-label use of an AI system, as well as reinforce the principle of responsibility. In short, DoD will need to rethink how it traces its AI systems, who has access to particular datasets and models, and whether those individuals are reusing them for other application areas. (Defense Innovation Board 2019, 35)

Traceability without transparency will be very limited. It may foster responsible uses of AI and compensates in part for the lack of transparency of this technology, but does not allow for overcoming the risks related to the lack of transparency. Consider, for example, the predictability problem. Insofar as AI systems are black box, it is hard to foresee their behaviours in specific contexts of deployment and to scrutinise outcomes when these are unwanted or unintended. It is problematic that the US documents focus only on traceability, avoid discussion of the lack of transparency of many AI systems, and do not offer any suggestions relevant to mitigate the lack of transparency of AI systems, for example by prescribing assessment of the transparency of different AI systems and choice of the most transparent one available for a given task.

It is also worth stressing that traceability and transparency are an *infraethical* element (Floridi 2017); that is, their value is determined with respect to their impact on ethical principles and rights. This is why a focus on traceability or transparency needs to be specified with respect to scrutiny and responsibility. Without this reference, it is not clear what information should be disclosed—and to whom—and thus what records should be kept. When stated in generic terms, transparency and traceability risk favouring malpractices (Floridi 2019), like "ethics shopping (i.e., combining existing ethical principles, guidelines, codes, etc. to justify preexisting behaviours); blue-washing (i.e., making unsubstantiated or misleading claims about one's commitment to AI ethics), or implementing superficially AI ethics measures; lobbying (i.e., leveraging AI ethics principles to delay or avoid good and necessary relevant legislation); or ethics shirking (i.e., making unsubstantiated or misleading claims about, or implementing superficial measures in favour of, the ethical values) (Floridi 2019, 186).

2.4. Reliable and Governable

The DoD principle focusing on reliability of AI states that

> DoD AI systems should have an explicit, well-defined domain of use, and the safety, security, and robustness of such systems should be tested and assured across their entire life cycle within that domain of use. (Defense Innovation Board 2019, 8)

The supporting document stresses the need for reliable (rather than trustworthy) AI, whose "safety, security, and robustness . . . should be tested and assured" (Defense Innovation Board 2019, 8). This principle is specifically oriented around fostering verification and validation and improving AI robustness. This is a crucial requirement for the use of AI in defence, and one that is important to mention explicitly in order to reiterate the need to monitor AI systems, especially when these are deployed within a defence organisation (more on this in section 4.5).

At the same time, the supporting document highlights the importance of human agents being able to disengage or deactivate systems that show unintended escalatory behaviour. It also emphasises the need for human control given the unpredictable behaviour of some AI systems, especially those operating in complex and dynamic environments (Defense Innovation Board 2019, 39). Control is not mentioned explicitly in the DoD's principles, but it is central to the principle focusing on governable AI, which prescribes that

> DoD AI systems should be designed and engineered to fulfill their intended function while possessing the ability to detect and avoid unintended harm or disruption, and for human or automated disengagement or deactivation of deployed systems that demonstrate unintended escalatory or other behaviour. (Defense Innovation Board 2019, 4)

While pointing in the correct direction, insofar as it specifies the need to maintain AI under some form of control, the supporting document remains vague with respect to what the desirable forms of control should be, how this should be exerted, and what the minimum level of ethically acceptable control is. This is understandable, given that the idea of control of AI remains a contested one (Tsamados and Taddeo 2023). However, the lack of any suggestions as to how to operationalise control of AI in defence is a missed opportunity. These need to include models of control, for example human control in/out/post the loop, but also modes of integration of AI in decision-making processes, for example by specifying protocols that ensure that human agents remain in charge of the final decision in teams that include AI systems. To this end, a notable omission in the DoD's principles is the lack of focus on human autonomy, which enables stronger forms of control over the use of AI as it protects the ability of human agents to dissent from AI-based decisions and override them, whenever these are considered to be mistaken or inappropriate.

3. From Defence Principles to Practice

In 2022 the US DoD published the "Responsible Artificial Intelligence Implementation Pathway" (DoD Responsible AI Working Council 2022). The document outlines an implementation pathway for the AI ethics principles it published in 2020. It builds on the DoD's existing infrastructure for technology development and governance, including software engineering and robust data management practices, and provide an "enterprise-wide approach", defining responsibilities for stakeholders across the DoD.

The operationalisation of DoD's principles is structured around six tenets: RAI governance; war fighter trust; AI product and acquisition life cycle; requirements validation; RAI ecosystem; and AI workforce. The DoD states that implementing the RAI will require a flexible approach to deal with different needs and complexities, based on factors such as technical maturity and different use contexts. Product needs will differ throughout the life cycle of an AI system, and the DoD will need to find the right balance between responsibility, speed, and ease of implementing the RAI, while removing barriers to adoption and innovation (2022, 14). To implement the six tenets, the DoD sets out "lines of effort", which direct actions to implement best practices and standards, tasking the DoD with developing new approaches where necessary. The lines of efforts are accompanied by overarching goals, identification of those responsible for implementation, and estimated timelines for implementation.

The six tenets provide some guidance as to what goals should be kept in mind while operationalising the DoD's AI ethics principles, and delineate the institution's attitude towards the adoption of AI, but they do not offer any specific guidance to address the problems that may emerge when applying the principles to specific cases. For example, no guidance is offered to those responsible for the operationalisation of the principle as to how to balance ethical principles when they conflict with each other.

The US Defence Innovation Unit (DIU) (Dunnmon et al. 2021) proposes an erotetic approach to close this gap. They suggest a series of questions to provide step-by-step guidance for DoD stakeholders, including AI vendors and programme managers. The questions should facilitate the alignment of AI programmes with DoD ethical principles, while also ensuring that fairness, accountability, and transparency are considered at each step of the development cycle. The resulting RAI guidelines are set out in a graphical workflow that enables actors to consider specific questions at each phase of the AI life cycle. For each phase, the document offers a worksheet that serves as a documentation and verification mechanism across planning, development, and deployment phases. For example, the planning phase requires personnel from the government agency requesting the system to collaborate with the programme manager to define the system's functionality, the resources required, and the operational context, in accordance with the DoD's ethical principles (Dunnmon et al. 2021, 8).

The DIU erotetic approach builds on the RRI approach, and inherits its strengths and limitations. As mentioned in section 1, the RRI approach aims to foster responsible choices through anticipating and gaining knowledge of possible consequences of research and innovation, and building capacity to respond to them. This approach rests on four elements: anticipation, reflexivity, inclusion, and responsiveness (Stilgoe et al. 2013). Anticipation involves systematic thinking aimed at increasing resilience while revealing new opportunities for innovation and risk research. Reflexivity entails an assessment of one's activities, commitments, and assumptions. Inclusion refers to the involvement of new voices in the governance of science and innovation to increase legitimacy. Responsiveness refers to the capacity to change the shape or direction of innovation in response to changing public values and circumstances. The four elements shape lines of questioning around the processes and purposes of innovation, but while RRI approach was conceived to foster critical reflection in researchers with respect to the societal impact of their research, it does not offer specific guidance to address ethical risks in specific scenarios. This is why it is not adequate when applied to high-risk domains, where concrete guidance and critical reflection are needed. Albeit welcome, the RRI approach is insufficient to ensure that the adoption of AI adheres with democratic values and domain-dependent ethical principles.

The RRI approach helps avoid the risk of transforming AI ethical compliance into a box-ticking exercise, insofar as it fosters critical reflection on the ethical implications of AI. However, if not coupled with institutional mechanisms to interpret AI ethics principles and operationalise them, it leaves the decision as to what is ethically acceptable entirely to local decision-makers (researchers, developers, operators, practitioners, etc.) who may have little or no expertise in AI ethics to make these decisions. This poses two risks. The first risk is that resulting decisions as to how to identify and mitigate ethical risks or how to define ethically acceptable outcomes are trivialised and reduced to commonsensical outcomes. The other risk is that normative decisions may be made without any normative justification (Kim et al. 2009) by actors without a normative authority (e.g., practitioners with no relevant expertise in ethics; actors operating without a clear and scrutinisable methodology; or decisions made without considering relevant stakeholders), thus opening questions about their legitimacy. This is problematic when applied to the high-risk cases typically found in defence, where decisions concerning the use of AI may impact individual rights and democratic values and involve risks related to the use of force.

When applied to high-risk domains, like defence, the RRI approach leads to *ethics devolution*. This is a malpractice which occurs when the burden of interpreting AI ethics principles, identifying criteria to harmonise conflicting principles, and implementing them in specific contexts is shifted from institutions to their members (employees), who may lack the necessary expertise, resources, and understanding of ethical risks to make normative decisions.

Ethics devolution can lead to oversimplification of this type of ethical challenges, for example when these are reduced to health and safety issues, and trivialisation of ethical solutions and malpractices, like ethics blue-washing and ethics shrinking (Floridi 2019).

Consider, for example, the method proposed in the DIU's document. It embeds normative elements in the questions— "have you *clearly* defined tasks?"—without giving guidance as to how to address these elements. The level of clarity required when defining tasks and who is responsible for setting this threshold are both left unspecified. In the same way, a question like "are end users, stakeholders and responsible mission owners identified?" presupposes the specification of the criteria and procedure for identifying stakeholders and their interests. The debate on the identification of stakeholders is wide (Donaldson and Preston 1995; Seppälä, Birkstedt, and Mäntymäki 2021; Ayling and Chapman 2022; Georgieva et al. 2022). This why it is problematic to ask practitioners whether stakeholders have been identified without providing a method to identify them. Specifying this method is not a trivial task and has severe normative implications, but without it answers to these type of questions can only remain vague and unsatisfactory.

The implementation of AI ethics principles is a normative act itself. Among other things, it entails decisions with respect to balancing conflicting interests, definition of risk thresholds, and an understanding of what is socially acceptable. For this reason, it can neither be left to the critical reflection of individuals nor reduced to internal policies of an institution. Instead, it has to be conducted following a reproducible (and therefore scrutable) methodology, leveraging relevant expertise and ensuring independence of those implementing the principles, as well as representation of all the involved stakeholders. I shall return to these points in section 5, after offering five high-level ethical principles for AI in the defence domain that have been specified, building on the lessons learned from the analysis of the DoD's principles and avoiding the limitations outlined in section 3.

4. Five Ethical Principles for AI in Defence

The principles I propose in this section have been designed to address specific ethical problems posed by the deployment of AI in the defence domain.

The five principles are these:

1. Justified and overridable uses
2. Just and transparent systems and processes
3. Human moral responsibility
4. Meaningful human control
5. Reliable AI systems

I describe each of these in turn in the next sections.

4.1. Justified and Overridable Uses

The principle of justified and overridable uses states that

> the adoption (or not) of AI needs to be justified to ensure
> that AI solutions are not being underused, thus creating op-
> portunity costs; or overused and misused, thus creating risks.
> Similarly, the decision whether to resort to AI should always be
> overridable, in the event of unwanted consequences.

Even when designed and deployed according to ethical principles,
AI remains an ethically challenging technology. Its use could lead
to great advantages for national defence, but it is not a silver bullet.
As has already been noted by Floridi and colleagues:

> it is important to acknowledge at the outset that there are
> myriad circumstances in which AI will not be the most effec-
> tive way to address a particular social problem. This could be
> due to the existence of alternative approaches that are more ef-
> ficacious or because of the unacceptable risks that the deploy-
> ment of AI would introduce. (Floridi et al. 2020, 1773)

Some of these risks include AI encroaching on human rights or
IHL or posing risks to international stability (the reader will recall
the risks of the snowball effect linked to the adversarial and non-
kinetic use of AI mentioned in Chapter 1). This is why the decision
whether or not to use AI systems should follow a careful analysis of
the ethical risks and benefits in any given context of deployment.
It is problematic that the US DoD's, but also the UK MoD's and the
NATO's, principles do not address this point expressly, prescribing
a cost-benefit analysis of ethical risks and opportunities of using
AI in defence as a necessary condition for the decision to use this
technology.

This principle yields different recommendations depending on the precise use to which AI is put. When considering the use of AI for sustainment and support, the principle calls for a weighing of the benefits of using an AI system to speed up a decision-making process or optimise resource logistics and distribution against the possibility that it could have a negative impact on human expertise's and autonomy or on individual and group rights.

When deciding whether or not to deploy AI for adversarial, whether kinetic or non-kinetic, purposes, it is essential to ensure that AI systems will respect the Just War Theory principles of necessity, distinction, and proportionality ("The UK and International Humanitarian Law 2018" n.d.). As we shall see in Chapter 4, this may prove to be a hard task. Consider adversarial and non-kinetic uses of AI: the principles of Just War Theory are geared toward kinetic forms of conflict, and therefore their application to non-kinetic warfare may not be obvious. For example, proportionality entails weighing the expected damage to intangible entities (e.g., data or services) against the concrete military aim to be achieved (Taddeo 2012a, 2012b, 2014a). Respecting this principle will require extending the scope of the fundamental tenets of Just War Theory from kinetic to non-kinetic operations, a hard but not impossible task.

4.2. Just and Transparent Systems and Processes

The principle of just and transparent system and processes prescribes that

> AI systems (or their use) ought not to lead to any breach of the principles of Just War Theory nor should they perpetrate any undue discrimination and should be as transparent as possible to perform effectively their task.

To respect this principle, AI defence institutions should ensure that the deployed AI systems, and the processes in which they are embedded, remain traced and explicable to facilitate the identification of the origin of any breach of the principles of Just War Theory, of unintended and mistaken outcomes, and of accountability, and to guarantee the possibility of scrutinising and challenging processes and outcomes to ensure that they remain ethically sound.

As discussed when considering the US DoD's principles, it is crucial to maintain the principle of justice and transparency in the correct relation, whereby the latter is an enabler of, and not a substitute for, the former.

To respect this principle, four aspects are crucial:

- ensure that the entire AI is used in respect of the principles of Just War Theory and its principles inform the entire life cycle of these technologies;
- maintain traceability for the design, development, procurement, and deployment of AI systems;
- set standards for transparency levels and traceability of processes;
- establish processes for ethics-based auditing involving the AI lifecycle and the entire decision-making process in which AI is included, to ensure that both human and artificial agents understand, follow, and respect the relevant ethical principles (Mökander and Floridi 2021a).

Defence agencies should participate actively in the life cycle of AI technologies that they procure, and inform the design and the development phases by setting standards for transparency and traceability, while also offering a trusted space where these technologies could be beta-tested. To facilitate this process, procurement policies should involve ethical scrutiny of the involved third parties. National interest and security make it likely that scrutiny in this area may not be public; nonetheless it is important that scrutiny is conducted by independent bodies or committees, which should be enabled and supported to develop an objective, in-depth assessment.

4.3. Human Moral Responsibility

The principle focusing on human moral responsibility mandates that

> humans are the only agents morally responsible for the outcomes of AI systems used for defence purposes.

Respecting this principle proves to be problematic, due to the distributed and interconnected ways in which AI is developed and the lack of transparency and predictability of its outcomes. A key problem here is the gap between the intentions of the human agents involved in the AI life ycle and the behaviour of the AI systems once put in use. This creates a responsibility gap; that is, there is a set of behaviours of AI systems that cannot be connected causally to the actions and intentions of human agents and for which humans cannot be held morally responsible. Three approaches have been suggested in the literature to close this gap:

- following the chain of command, control and communication (a linear approach)
- a faultless, back-propagation approach (a radial approach)
- voluntary acceptance of moral responsibility ("the moral gambit").

I will analyse the voluntary acceptance of moral responsibility in Chapter 7. Here, I will focus on the first two approaches. They can be described more simply as a linear and a radial approach, respectively. They are complementary, in that they serve the twin purposes of addressing unwanted consequences, misuses, and overuses of AI, and aim to foster a self-improving dynamic in the network of agents involved in the design, development, and deployment of AI for defence.

According to the linear approach, responsibility follows the chain of command, control, and communication. In this case, decision-makers are held responsible for the unwanted consequences of AI, whether they result from failures of AI systems, unpredictability of outcomes, or bad decisions. In order to ascribe responsibility fairly, it is essential that decision-makers have adequate information and understanding of the way the specific AI system works in any given context, of its robustness, of the risks of unpredicted (and unwanted) outcomes, of the required level of meaningful control, and of the dangers that may follow if an AI system fails to behave according to expectations. The linear approach entails a certain epistemic threshold. This means that the use of AI must be coupled with proper training of personnel, including those who decide to deploy AI systems and those who use it, so that they understand the ways in which AI systems work, the risks and benefits linked to the systems, and the ethical and legal implications of their decisions. This approach rests on the idea that informed decision-makers choosing to use AI do so while being aware of the risks that this may imply and that full responsibility for the choice is attributed to them.

The radial approach is useful to address unwanted outcomes of AI systems that do not stem from the intentions of human agents or that follow from actions that are morally neutral *per se*. Accordingly, this approach addresses the ethical consequences arising from the convergence of different, independent, morally neutral factors. This has been defined as "faultless responsibility" (Floridi 2016), referring to contexts in which, while it is possible to identify the causal chain of agents and actions that led to a morally good/bad outcome, it is not possible to attribute intent to perform morally good/bad actions to any of those agents individually. In this case, all the agents are held morally responsible for that outcome insofar as they are part of the network that determined it.

This approach is akin to the legal concept of strict liability. According to strict liability, legal responsibility for unwanted outcomes is attributed to one or more agents for the damage caused by their actions or omissions, irrespective of the intentionality of the action and feasibility of control. When considering human-machine teaming—that is, the integration of AI systems in defence infrastructures, decision-making processes, and operations—what one needs to show to attribute moral responsibility according to the radial approach is that

> some evil has occurred in the system, and that the actions in question caused such evil, but it is not necessary to show exactly whether the agents/sources of such actions were careless, or whether they did not intend to cause them. (Floridi 2016, 8)

All the agents of the network are then held maximally responsible for the outcome of the network. As Floridi (2016) stresses, this approach does not aim at distributing reward and punishment for the actions of a system; rather it aims at establishing a feedback mechanism that incentivises all the agents in the network to improve their outcomes—if all the agents are morally responsible, they may become more cautious and careful and this may reduce the risk of unwanted outcomes. This becomes quite effective when, for example, the moral responsibility is linked to the reputation of the agents.

When combined, the linear and the radial approaches can help to close the responsibility gap. Nonetheless, the linear approach offers a limited solution insofar as it ascribes responsibilities on the basis of the line of command, which is not sufficient to satisfy the criteria of intentionality and causal connection required to ascribed moral responsibility in a fair and justified way. The radial approach fosters responsible behaviour, but does not help in ascribing moral blame or praise for the outcomes of AI systems in defence. This is a crucial gap, for the use of AI in defence may lead to severe breaches of principles, values, and rights, and attributing moral responsibility is a necessary condition for maintaining the morality of warfare in the digital age. I shall return to this point and offer a solution to close the responsibility gap in Chapter 7.

4.4. Meaningful Human Control

The concept of meaningful control has been discussed widely in the literature on LAWS, and indeed when considering these systems, control is a key element to consider. However, meaningful control is necessary also when considering uses of AI that may not lead to the use of force. This is because

> military systems must be able to function safely and effectively under a wide range of highly dynamic environments and use cases that are hard to predict or anticipate during the design phase. They must also be resilient to failure and to complex, uncertain and unpredictable events and situations where the dynamics of the military domain necessitate complex judgements regarding acceptable actions based on rules of engagement, international law and judgements over legality, proportionality and risk. (Boardman and Butcher 2019, 2)

Thus the principle prescribing meaningful human control mandates that

> the entire AI life cycle ought be under meaningful forms of human control, in order to limit the risk of AI systems not meeting the original intent, to identify any mistakes and unintended consequences, and to ensure timely intervention should this become necessary.

Meaningful human control of AI is characterised as dynamic, multidimensional, and situation dependent, and it can be exercised through different elements of a human-machine teaming. For example, the Stockholm International Peace Research Institute and the International Committee of the Red Cross identify three main areas of human control of weapon systems: the weapon system's parameters of use, the environment in which it is deployed, and human-machine interaction (Boulanin et al. 2020). More parameters can also be considered. For example, Boardman and Butcher (2019) suggest that control should not just be meaningful but also "appropriate", insofar as it should ensure that human involvement in the decision-making process remains significant without impairing system performance.

As it might be expected, there is a lack of consensus on what is meant exactly by *meaningful* control, but there are thresholds below which control is so minimal as to become irrelevant and above which control makes the use of the system inefficient. Hence, the principle can be implemented minimally and maximally. Minimally, the implementation of this principle requires having a human *on* the loop who is able to understand the functioning of the system and its implications and with the ability to unplug the system in a timely and effective manner. Maximally, the principle requires individuals in charge of AI systems to combine technical, legal, and ethical training to ensure that the decision *to let the system work* is informed by all relevant dimensions, and not a mere vetting of the system.

In this regard, the principle does not admit fire and forget uses of AI, as it prescribes control as an element to be modulated according to a rigorous risk assessment of both unintended consequences and any negative impact on the principle of Just War Theory, national defence, and international stability. Where meaningful control cannot be undertaken with these assessments, the use of AI systems is ethically unwarranted. It should be noted that the principle of meaningful human control is best implemented when protocols for the attribution of moral responsibility for unintended outcomes, misuses and mistakes made by AI systems are in place, alongside effective redressing and remedy processes.

4.5. Reliable AI Systems

Both the US DoD's and the UK MoD's sets of principles include reliability. This is crucial. Fostering reliability means facilitating control of the outcomes of a systems as well as the security of both users and infrastructures relying on AI. Reliability of AI implies robustness of AI technologies. However, as we saw in Chapter 1, AI has a poor shock response (robustness) and any slight alterations to inputs can degrade a model disproportionately (Rigaki 2017). Thus, deploying AI for defence purposes could favour opponents (Brundage et al. 2018; Taddeo, McCutcheon, and Floridi 2019) if the system is not deployed according to procedures that envisage monitoring and prompt intervention in the case of mistakes or system degradation. This is why this principle prescribes monitoring of AI systems throughout their deployment, as well as putting in place measures to verify and validate the systems and assess their robustness. The principle of reliable AI systems prescribes that

> there ought to meaningful forms of monitoring of the execution of the tasks delegated to AI and of the entire life cycle of the system as well as clearly defined risk threshold related to systems' robustness and predictability. These should be adequate to the learning nature of the systems, and their lack of transparency, and their purpose of use while remaining feasible in terms of resources, especially time.

Monitoring may include new forms of procurement that envisage an active role of the defence institution in the design and development process; in-house design and development of AI models; data for system training and testing being collected, curated, and validated directly by the systems providers and maintained securely; mandatory forms of adversarial training with appropriate levels of refinement of AI models to test their robustness; sparring training of AI models; and monitoring the output of AI systems deployed in the wild with some form of *in silico* baseline model, as suggested by Taddeo, McCutcheon, and Floridi (2019).

As discussed in Chapter 1, AI systems are autonomous, self-learning agents that interact with their environment. Their behaviour depends as much on the inputs they are fed and interactions with other agents once deployed as it does on their design and training. To be ethically sound, or simply to mitigate ethical risks, uses of AI for defence purposes need to take into account the autonomous, dynamic, and self-learning nature, and therefore the limited predictability, of AI systems, and start envisaging forms of monitoring that span the entire life cycle to limit the risks of unintended and unwanted outcomes.

5. A Three-Step Methodology to Extract Guidelines from AI Ethics Principles in Defence

Like the principles developed by the US DoD, the UK MoD, and NATO, the principles I describe are high-level principles. They need to be interpreted in order to extract specific ethical guidelines. Here, I propose a methodology for interpreting AI ethics principles in order to specify guidelines. The methodology is designed to meet the specific risks faced by public institutions working in high-risk domains—for example, national defence and security, healthcare, and administration of justice—when they interpret AI ethics principles. Three categories of risks and problems are particularly relevant here: the moral legitimacy of the resulting guidelines; the risk of ethics devolution; and the reproducibility and scrutability of the process used to define the requirements.

To avoid these risks, I suggest that the interpretation of the principles be left to an independent, multistakeholder ethics board (EB), which follows a three-step methodology to extract ethical guidelines from high-level principles. The three steps involve abstraction, elicitation of ethical requirements, and harmonisation. The methodology is an iterative process that is refined through its implementation. It is important to stress that this process does not happen in a vacuum, but is influenced by, and strives to be consistent with, the rights and values already underpinning the work of defence institutions in democratic societies (see Figure 2.1). Let me first outline the EB, before delving in each of the three steps.

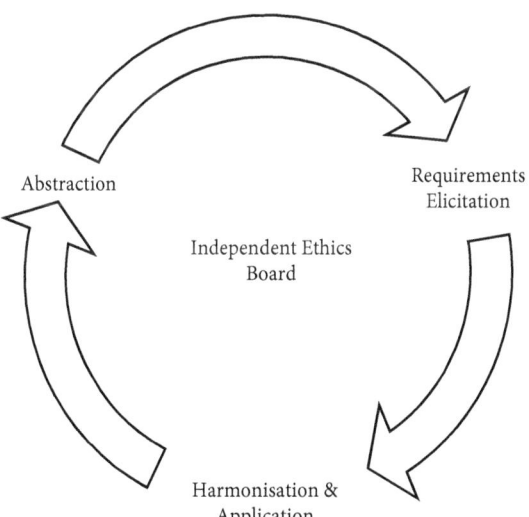

Figure 2.1 The three-step methodology for the definition of ethical guidelines for AI in defence.

5.1. Independent, Multistakeholder Ethics Board

The EB has three tasks: identify the LoA to model the AI life cycle; interpret AI principles in order to extract specific requirements to be met at each step of the AI life cycle; and define the criteria to inform purpose and context-specific balancing of the principles. These tasks require in-depth expertise in AI, AI ethics, military ethics, Just War Theory, national defence and security, as well as expertise in civil rights, democratic values, international humanitarian laws, and the internal procedures of the relevant defence institution. Thus, the EB should include experts in all these areas. Aside from the breadth of expertise, the EB must be independent of the institution adopting the principles and the resulting guidelines and should include representatives of the different categories of stakeholders impacted by the use of AI in defence. These comprise, for example, representative of the military and non-governmental organisations representing civilians. For the EB to be effective, it is crucial that all relevant stakeholders are involved, but also that they take an active role in the shaping of the guidelines.

Davies, Ives, and Dunn (2015) categorise stakeholders' involvement in the specification of ethical guidelines into two approaches: dialogical or consultative. Under dialogical approaches, ethical analysis forms part of the stakeholder engagement process itself. In this case, consensus-based methods justify normative conclusions (Widdershoven, Abma, and Molewijk 2009). Some dialogical approaches rely on the idea that dialogue can lead to individuals reaching a shared understanding of the world, leading to agreement on the correct solution. Other interpretations argue that democratic authority provides normative justification rather than shared interpretation and consensus (Kim et al. 2009). In this case, justification flows from the legitimacy of the process used to set up the board and to come to conclusions, rather than the actual outcome or solution. Under consultative approaches, ethical analysis is undertaken after explicit engagement with stakeholders, for example, via a workshop, focus group, or deliberative mini-public. The views of the stakeholders feed into the ethical analysis, but they are not involved in it directly. With consultative approaches, results are justified on the basis of the coherence of the proposed solutions with the adopted moral theory (Davies, Ives, and Dunn 2015).

Both approaches offer important insights when considering the definition of ethical guidelines in the defence domain. For example, the discursive element of the dialogical approach and the need to develop consensus around ethical risks and desirable solutions is key when developing ethical guidelines for AI systems that will impact different stakeholders differently. At the same time, the debate on reflexive balancing—that is, balancing competing ethical principles in specific contexts—which is central to consultative approaches, is also very relevant when considering the defence domain.

Adopting the distinction proposed by Davies and colleagues, the EB will work best if constructed following a revised version of the dialogical approach. An appropriate representation of all legitimate interests, and consensus reached through dialogue, transparency, and independency of process will provide the normative justification of the decision of the EB.

The EB should aim to achieve a *moral impartiality* (Habermas 1990) and to find fair ways to reconcile different interests (McCarthy 1995; J. Heath 2014); that is, the board should produce recommendations with consequences that can be accepted as fair by all involved parties. A multistakeholder approach and the independence of the EB are essential to this end, to ensure that the interests of those impacted by AI in defence are represented and respected adequately.

To achieve moral impartiality, the EB should work following Habermasian theory of discourse ethics (Habermas 1990, 1998, 2021). According to this theory, the involved stakeholders commit to consider rationally the interests of the others, argue for their own positions, and work to find universally agreeable norms to resolve a specific conflict of interests by identifying norms whose effect could be accepted by all parties. This will be particularly relevant when the EB has to give indications as to how to balance ethical principles in specific contexts (more on this in section 5.3).

Three reasons support this multistakeholder approach and the independence of the EB. The first is that an independent EB would avoid risks of manipulation. Evidence of manipulation of attempts to interpret ethical principles has been reported in the literature (M. Krishnan 2020; Fazelpour and Lipton 2020; Terzis 2020). Indeed, as Morley et al. stress:

> AI practitioners may choose the translational tool that aligns with what is for them the most convenient epistemological understanding of an ethical principle, rather than the one that aligns with society's preferred understanding. (2021, 243)

This creates risks of ethics shopping and ethics washing (Floridi 2019), as well as the risk of reputational damage linked to these malpractices. The second reason to support a multistakeholder approach is to avoid the risk of ethics devolution, which could lead personnel to consider efforts to develop and apply ethical guidelines as a mere add-on or an extra burden. This, in turn, would undermine both the development of a pro-ethical institutional culture and the outcomes of any AI ethics initiative. Third, an independent board could foster a genuine pro-ethical attitude within a defence organisation, whose members would perceive that the institutional focus on ethics is not superficial and that they can rely on independent expertise to identify and mitigate ethical risks.

The time has come to focus on abstraction, the first step of the methodology.

5.2. Abstraction

The consensus in the literature is that AI ethical guidelines should span the entire life cycle of an AI system (Alshammari and Simpson 2017; d'Aquin et al. 2018; US Department of Defense 2022b; Cihon, Schuett, and Baum 2021; Dunnmon et al. 2021; High-Level Expert Group on Artificial Intelligence 2019; Ayling and Chapman 2022; Mäntymäki et al. 2022). This emphasis on a life cycle approach tallies with a broader consensus that AI ethics governance must be systemic to be effective (Eitel-Porter 2021). The focus on the life cycle also mandates the iterative (re)application of principles at successive stages of the project. This is important from a process, product, and purpose point of view (Stilgoe, Owen, and Macnaghten 2013). Regarding process, the needs of a particular project are likely to evolve beyond those envisaged at the beginning, and with them new ethical risks may emerge. From a product point of view, some AI models, like generative models, can produce unexpected behaviours (Taddeo et al. 2022), and therefore ensuring that the product continues to adhere to ethics principles beyond its release is essential. From a purpose point of view, the social and political motivations of a project and the goals of innovation may change over time, thus ensuring control over the project requires continuous monitoring of its ethical implications.

It is important to remark that, insofar as the AI life cycle is a model of the processes and conditions of design, development and deployment of AI technologies, it carries a normative value. Those who define the AI life cycle define the scope of applicability of both the principles and the guidelines. The life cycle of AI is a socio-technical process where a

> neat theoretical distinction between different stages of technological innovation does not always exist in practice. (La Fors, Custers, and Keymolen 2019, 210)

This can make it difficult to identify the points at which ethical questions should be asked, particular steps taken, and specific goals met. Defining the right level of granularity in the description of the life cycle (and of the LoAs) can also be difficult. If too few stages are identified, gaps and blind spots might lead to ethical risks. If too many tasks are identified, the iterative application of the principles multiplies, making the guidelines unwieldy and quickly outdated by rapid developments in the technology.

When considering the AI life cycle, the correct LoA is the one that allows for identifying the steps in which ethical risks might emerge and also for identifying the actors accountable for these steps. This is why I combine three LoAs focusing on steps of the AI life cycle (LoA_{Steps}), the actors accountable for those steps (LoA_{Actors}), and the ethical risks that can emerge at those steps (LoA_{Risks}). An AI life cycle can be described in terms of stages—for example, design, development, and deployment; steps compounded in each stage; or tasks entailed by each step. The focus on steps of the AI life cycle avoids the risks of developing a model that is either too generic (when focusing on stages) or too meticulous (when focusing on tasks). As for LoA_{Actors}, I include both those who provide the technology or contribute to its design and development (i.e., providers) and those who decide on and monitor the use of an AI system (i.e., users). Both companies providing the technologies and defence organisations developing or using them have internal hierarchies and structures that shape the ways in which different parts of the organisation become involved in the AI life cycle. Thus, I suggest that actors accountable for the implementation of the guidelines for the different steps are identified following existing structures. Below, I offer an example of how a model of the AI life cycle would look given the proposed LoAs. It is one of the tasks of the EB to identify the correct LoA and specify ethical risks accordingly. The one below is simply an outline of a possible model.

The model results from adapting the AI life cycle model proposed by Floridi et al. (2022) as shown in Figure 2.2, which rests on widely

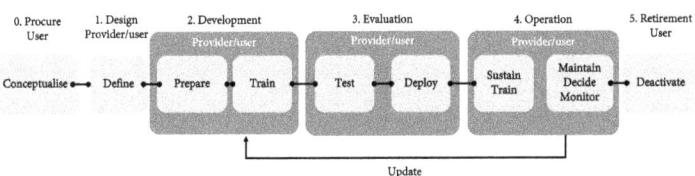

Figure 2.2 A model of AI life cycle model adapted from Floridi et al. 2022.

adopted standards for software development such as ISO, IEC, and IEEE standards (ISO/IEC TR 24748-1 and ISO/IEC/IEEE 12207:2017).[4] Given that the model has also been proposed to define a process for ethics-based auditing of AI (Floridi et al. 2022), it already includes ethically relevant aspects. For example, it includes an evaluation stage, which is relevant when considering high-risk applications of AI such as those in the defence domain. The original model includes five stages—that is, design, development, evaluation, operation, and retirement. It has been adapted here to include a procurement stage, to identify those early points in an AI life cycle where decisions can lead to unethical consequences. The procurement and design stages are distinct, in that the former includes the specification of a possible use case, consideration of its utility and impact, and requirements for providers, while the latter refers to the definition of specific technical requirements, for example, data needs and architecture of the AI model, which will inform the actual end system to be used.

The AI life cycle model resulting from the adaptation of the model provided by Floridi et al. (2022) is shown below in Table 2.2. Note that the model could be expanded to consider specific steps

---

[4] https://www.iso.org/standard/72896.html and https://www.iso.org/standard/63712.html. It is worth noticing that standards to model the AI life cycle are now emerging; see, for example, the ISO/IEC DIS5338 (https://www.iso.org/standard/81118.html). An EB could start from here to design an ethics model of the life cycle of an AI system.

Table 2.2 The model of the AI life cycle developed using LoA_{Steps}, LoA_{Actors}, and LoA_{Risks}

Life cycle stage	Accountable actor	Steps	Example of ethical risks
Procurement	User	• Conceptualise use case, context, architecture, and objective • Specify user requirements • Specify concept of employment • Assess fitness of the AI solution for the problem • Request for provider's qualifications	• Disproportionate solution • Lack of transparency of the AI model • Responsibility and accountability gap • Lack of transparency and traceability from the provider
Design	User/provider	• Define data needs • Define trade-offs of algorithmic decisions • Provide risk analysis and define risk thresholds	• Limited robustness of the model • Responsibility gap • Disproportionate data collection (privacy breaches) • Transparency/efficiency balance inadequate • Design assumptions do not account adequately for contextual factors (e.g., racism, economic factors, complex environment)
Development	User/provider	• Sourcing data • Data analysis • Preparing data • Splitting data • Build and train an initial model • Develop a benchmark	• Data collected without proper consent • Model drift

Table 2.2 Continued

Life cycle stage	Accountable actor	Steps	Example of ethical risks
Evaluation	User/provider	• Test for undue outcomes, e.g., bias • Test for robustness • Evaluate primary metrics • Refine the model • Select deployment strategy	• Undue discrimination • Limited predictability • Accountability and responsibility gap • Vague specifications for deployment leading to unforeseen outcomes
Operation	User	• Monitor and trace • Post-deployment review • Define accountability • Establish feedback mechanism	• Accountability gap • Improper use leading to unforeseen outcomes • Communication flaws leading to accountability gaps
Retirement	User	• Assess deactivation risks • Archive logs	• Lack of logs

related to the way a project is handled and deployed in a defence organisation. For example, an EB working in the UK could consider expanding the model to include specific steps of the acquisition cycle (CADMID—Concept, Assessment, Demonstration, Manufacture, In-Service, Disposal),[5] whenever these are deemed to introduce specific ethics risks.

---

[5] https://www.rpsgroup.com/cadmid-cycle/#:~:text=What%20is%20the%20CAD MID%20cycle,and%20Disposal%20(CADMID)%20cycle.

5.3. Interpretation and Requirements Elicitation

AI ethics principles have been compared to constitutional principles (Morley, Floridi, et al. 2020). Like constitutional principles, ethical principles are meant to be foundational rather than
offering detailed guidelines. They embed values more than specific directives, are expressed in simple and plain language, and
have an "open-textured character [and a] purpose-oriented nature" (Dehousse 1998, 76). They are often articulated in a non-
hierarchical way, but they are competitive and may need to be
balanced against each other depending on the specific context of application. These shared characteristics make judiciary
methodologies used to interpret constitutional principles effective
in aiding the interpretation of ethics principles as well. This is why
I look to the teleological methodology used for the interpretation of
constitutional principles by constitutional courts of justice to consider how to interpret AI ethics principles across the life cycle of AI
technologies in the defence domain.

The literature identifies five methodologies with which a judge
may interpret constitutional principles (Llorens 1999): literal, historical, contextual, comparative, and teleological.[6] I will focus here
on the teleological methodology and disregard the others. This is
because literal and contextual methodologies focus on the exact
meaning of the words of the principles and the immediate context
of application, respectively, to understand the prescription of principle, but they disregard their overall goal. Because of this, these two
methodologies can lead to "absurd" results when they lead to "to an
interpretation clearly contrary to the objective of the legislation in

---

[6] The literature on jurisprudence of the courts of justice is vast, with much discussion
on methodologies for judiciary interpretation, namely, the methods with which a judge
assesses cases as a matter of justice administration, division of powers in democracies,
and rule of law. We can disregard these issues when considering the interpretation of the
ethical principles, which are voluntary principles that organisations adopt without respect to legal requirement or compliance.

question" (Llorens 1999, 376). Historical methodologies focus on the intention of the legislator and/or the function of the principle at the moment of its ratification. Here, the intentions of the legislator (i.e., a parliament) are considered insofar as the legislator is a representative body expressing the will of the public, which is not necessarily the case when considering the defence institutions that have drafted AI ethics principles. In this case principles may result from the work civil servants not elected by the public. Thus, this methodology is not fit for the purpose of our analysis. The comparative methodology refers to the option to consider interpretation of similar principles adopted by other courts of justice. In the case of AI ethics principles, there is not yet an established tradition or approach to interpret the principles, so this methodology is currently unfeasible.

The teleological methodology focuses on the purpose underpinning the principles, on their context and goal, and their commitment to the effectiveness of the interpretation (*effet utile*). It rests on Article 31.1 of the Vienna Convention on the Law of Treaties 1969, which states that

> [a] treaty shall be interpreted in good faith in accordance with the ordinary meaning to be given to the terms of the Treaty in their context and in the light of its object and *purpose*.[7]

The teleological interpretation examines the words of a principle to identify its spirit, that is, the essential values and rights that the principle aims to protect. It considers context and objective, that is, the context in which a principle is stated, for example the specific treaty and the goals of that treaty. Decisions about how to clarify the purpose of a principle—and therefore how interpret it—can be aided by considering the documents that were used in its drafting.

---

[7] Vienna Convention on the Law of Treaties, opened for signature, May 23, 1969, 1155 U.N.T.S. 331, 8 I.L.M. 340, 8 I.L.M. at 691–92.

The methodology must be effective; that is, once the purpose of a principle is identified, it will be interpreted so as to achieve effectiveness, consistency, and uniformity with the legal framework of a given state (Fennelly 1997; Brittain 2016).

Following the teleological methodology, when considering AI ethics principles, an EB should first consider the spirit of an organisation's principles to identify the values and rights that they protect. Here the discourse ethics theory described in section 5.1 will be crucial to reconcile the views of the different stakeholders. The board should consider both the context and the objective. In the case of ethical principles issued by a ministry of defence, the context is quite straightforward to identify, given that they will refer to sets of other values, for example, democratic values, military ethics values, and the ethical principles offered by Just War Theory. The objective refers to the overall goal of the principles of identifying and mitigating ethical risks. Finally, respecting the commitment to effectiveness, the EB will have to define effective and applicable measures to ensure that the objective is met and the values or rights protected in the principles are not breached.

In terms of how an ethics board achieves this, it would not be sufficient for the board to define a set of questions that developers and users should ask themselves with respect to the impact of a specific step of the AI life cycle (see section 3). Instead, a board should define requirements that need to be met at each step of the AI life cycle. In other words, the EB will work to answer the question "What do I [provider/designer/developer/user] have to do to ensure that this step of the AI life cycle respect the AI ethics principles?"

For example, when interpreting the principle of "justified and over ridable uses" (as outlined in section 4), the EB could specify the following requirements for an AI system to move from the procurement to the design stage:

- there must be a detailed analysis that maps the ethical risks of the envisaged AI system for a specified goal, and specifying mitigation strategies for each risk as well as a demonstration of the effectiveness of these strategies;

- there must be a cost–benefit analysis showing the organisational value of the proposed solution which demonstrates both beneficial outcomes and proportionality of possible breaches of rights and values;
- the AI solution should include the specification of procedures to ensure human override of the system.

The organisation receiving the recommendations of the EB should consider the requirements to be necessary conditions for an AI system to move through its life cycle and to ensure that the necessary measures are put in place to mitigate the ethical risks identified by the board. While I do not focus in this book on the operationalisation and verification of such requirements (given that these will depend on specific internal policies and institutional organisations), it is worth stressing that any efforts to interpret ethical principles into practice will be futile if guidelines are not adopted and respected at the institutional level.

5.4. Balancing The Principles

The third and final step of the methodology concerns balancing the principles against each other. Also in this respect, ethical principles recall constitutional ones, which are often competitive. In this case, courts often establish a context-dependent hierarchy. As Alexy puts

> the proposition of finding a *"conditional precedence relation"* (Alexy 2002, 52): if the conditions x are given, [principle 1] prevails over [principle 2]; if the conditions y are given, [principle 2] prevails over [principle 1]. (Cited in Guastini 2019, 312; emphasis added)

The definition of the conditional precedence relation is key to a fair balancing of the principles, but this can be tricky, as it has to consider the institutional and cultural elements of an organisation, as well as ensure that the resulting balance leads to outcomes consistent with the overall ethos of the defence institution and with democratic values. Here, the independence and multistakeholder nature of the EB offers some assurance as to the fairness of the defined conditional precedence relations, while the teleological methodology offers guidance as to how to ponder and balance competing principles in specific cases.

One way in which the EB could provide effective guidance is by expressing the conditional precedence relation in terms of purpose- and context-specific tolerance thresholds, that is, how strictly a requirement needs to be met. The EB might set a much lower tolerance threshold (leaving little flexibility) for the satisfaction of the ethical requirements for kinetic and adversarial uses than for sustainment and support uses. A board might also set risk thresholds so that under a certain threshold of risk, a default set of ethical requirements is applied, and above a certain threshold the board will have to consider specific cases. For example, lack of transparency and traceability on the provider's side for an AI system to be used for sustainment and support or for an AI system to be used for adversarial and kinetic use may create the same risk of a responsibility gap. However, should the risk materialise, its impact would be much greater for adversarial and kinetic uses. This may guide the EB to set a much lower tolerance threshold (leaving little flexibility) for the satisfaction of the ethical requirement for the kinetic and adversarial uses than for sustainment and support uses.

As I have said, questions concerning the operationalisation of the methodology are outside the scope of this book. However, the methodology has been designed to be practical and agile, and to help an ethical board identify ethical risks and requirements focusing on *types* of AI systems and *purposes* of use. In this way, a board might not have to consider and specify ethical requirements for every new AI system to be procured, developed, designed by a defence organisation.

6. Conclusion

The ethical principles and the methodology need to be coupled to a pro-ethical institutional culture. This will mitigate the risk that ethics will be perceived, or treated, as an add-on or an extra burden for practitioners. Institutional pro-ethical attitude would foster ethics as a constitutive, unavoidable element of everyday practices, which leads to the achievement of positive results. This is especially the case when considering defence institutions or other public bodies working in high-risk domains.

Such a pro-ethical attitude needs to be fostered and demonstrated at an institutional level. Two ways are particularly important: ethics training and enforceability. The requirements specified by the EB are best applied when practitioners understand them. Indeed, ethical outcomes are fostered when practitioners are aware of the ethical risks, problems, complexities, and opportunities that come from AI, and which make adherence to specified ethical guidelines necessary. Thus, ethics training should be provided at the institutional level and made accessible (if not mandatory) to practitioners. At the same time, ethical guidelines that are not enforced will undermine any efforts to develop ethical uses of AI. To this end it is crucial that accountability is clearly established, for example via ethics-based auditing processes.

The debate on the ethical governance of AI in defence is still in its early days, despite the long tradition of military ethics and Just War Theory. This delay, while not desirable, comes with the advantage of building on the expertise and experience of AI ethical risks in other domains. At the same time, the need to catch up on ethical governance may lead to oversimplifying approaches, whereby, for example, AI ethics may be reduced to health and safety protocols, and ethical risks are framed simply as safety hazards. The application of ethical principles in high-risk domains is an iterative process, which has to account for the evolving nature of these technologies, changes in legitimate interests and risks, and changes in societal values and, hence, in what is considered socially acceptable. All this comes with costs, in terms of organisational resources, economic resources, and expertise. These costs are unavoidable and need to be shouldered by defence institutions if we are to leverage the potential of AI for defence while protecting the values of our societies.

3

Sustainment and Support Uses of AI in Defence: The Case of AI-Augmented Intelligence Analysis

1. Introduction

If Chapter 1 introduced this book by outlining its scope and the methodology, Chapter 2 worked as a spoiler—the reader already now knows the principles that should inform the use of AI in defence. However, two questions still need to be considered: whether those principles are appropriate, that is, they address the right ethical risks and opportunities posed by the uses of AI described in Chapter 1; and whether they are sufficient to address those challenges. The rest of this book addresses these two questions. I begin in this chapter by analysing some of the key ethical challenges related to the use of AI for intelligence analysis (AIA)—otherwise known as "AI-augmented intelligence"—and offer some policy recommendations to mitigate them.

AIA is one of the most relevant examples of uses of AI for sustainment and support purposes. In defence, it can be crucial to gain and maintain advantages over the opponent, insofar as it aids information asymmetry by enabling analysts to meet the deluge of data created by digital communications and facilitate their analysis. Consider, for example, the use of AI systems to validate information from multiple sources to detect and identify threats (such as military aircraft undertaking bombing runs) and to alert civilians in the affected areas (Hala Systems 2022; see also Freeman 2021).

The Ethics of Artificial Intelligence in Defence. Mariarosaria Taddeo, Oxford University Press.
© Oxford University Press 2024. DOI: 10.1093/oso/9780197745441.003.0003

The speed of processing required to generate alarms from these multiple sources would not be possible for a human analyst alone.

Post-9/11, there has been a growing demand for information about individuals (such as terrorists and international criminals) rather than states *per se*, and the growth of digital communications met this demand by supplying detailed information about individuals in ways previously considered impossible (Omand and Phythian 2018, 142). For the intelligence community this has pressing ramifications. The availability of structured and unstructured data is so extensive as to "overwhelm all previous forms of analytic tradecraft and pattern recognition" (Weinbaum and Shanahan 2018, 4). Without the tools and capacities to make sense of that data, nations will cede strategic advantage to adversaries. As the US Defense Innovation Board has noted: "Whoever amasses and organizes the most data—about ourselves as well as our adversaries—will sustain technological superiority. Failure to treat data as a strategic asset will cede precious time and space to competitors or adversaries" (Defense Innovation Board 2017, 3). As well as aiding the intelligence community to meet and leverage the data deluge, AIA can benefit the intelligence community in a number of other ways. For example, it can foster co-ordination and standardisation. Intelligence organisations are often quite fragmented (Zegart 2005).[1] This is in part an institutional problem: intelligence organisations are composed of large bureaucracies, all with different classification procedures, cultures, and different methods of gathering intelligence. Fragmentation is also a consequence of the very nature of intelligence gathering itself: intelligence organisations tailor intelligence to the needs of specific users. From this arises a trade-off between good intelligence for the customer and an intelligence system that can be

---

[1] For the most comprehensive discussion of this see the 2002 joint congressional report *Joint Inquiry into Intelligence Community Activities before and after the Terrorist Attacks of September 11, 2001* (US Senate Select Committee on Intelligence 2002).

coordinated effectively (Zegart 2022, 49). For AIA to be effective it is crucial to standardise intelligence data across different locations and agencies, to enable access to analysts where and when it is required. This is likely to lead to the standardisation of methods of data visualisation, the extraction of a specific piece of data from a source in one location for retrieval in another, and the simplification of such methods.

AIA may also alleviate the risks of "cognitive traps". These are biases that fit or distort reality to preconceived ideas or patterns. Intelligence organisations already employ methods and tools for addressing these biases. As far as AI is not perpetrating any bias itself (Yang et al. 2018; Tsamados et al. 2021a), its use could enable the identification of patterns or trends that may have been discounted or disregarded by humans.

AIA can also protect analysts from harmful imagery and material. This is pertinent in cases where AI is used for online safety purposes, for example identifying child sexual abuse imagery, but also to scrutinise images and data concerning war crimes. Running AI models across both the content and the metadata would prevent the analysts from having to examine (multiple times) imagery firsthand, thereby protecting analysts from "unnecessary exposure to traumatically disturbing material" (GCHQ 2021, 19).

It is not all positive, however, given that the use of AI for intelligence poses serious ethical risks. Some are similar to ethical risks arising from the use of AI in other domains, like undue discrimination, technology bias, lack of transparency; others are more domain-specific and refer, for example, to the risks of AIA fostering authoritarianism and political security, that is, the ability of governments to "take advantage of improved capacity to analyse human behaviours, moods, and beliefs on the basis of available data" (Brundage et al. 2018, 6) to identify and punish behaviours that contradict, contest, or merely do not align with governmental views or ideology.

I shall delve into these risks in the rest of this chapter. In particular, in section 2, I outline uses of AI for AIA with reference to current uses. In section 3 I map key ethical risks, analyse their implications, and offer policy recommendations to mitigate them. I conclude the analysis in section 4.

Before turning to AIA, I shall note two limitations of the proposed analysis. The first is that the uses of AI by the intelligence community are mostly secretive. This limits the extent to which existing use occurrences of AIA can be mapped. The second limitation follows from the first and concerns the findings of this chapter. While they address, and are relevant to, the use of AIA in general, the literature covered refers predominantly to intelligence activities by the US. This reflects the wider literature available on AIA employed by US intelligence agencies than other agencies. It should be noted however, that, being a leader in intelligence capabilities, where the US leads, international partners often follow. Focusing on the US thereby enables consideration of potential interoperability issues with other countries. Let me turn first to the uses of AI for intelligence in defence.

2. Mapping Augmented Intelligence Analysis in Defence

The concept of intelligence analysis is still contested in the literature (Ish, Ettinger, and Ferris 2021), with different authors and institutions providing different definitions, for example focusing on the socio-cognitive process (Johnston 2005, 37) and data analysis (Akhgar and Yates 2013, 181) to extract valuable information, as stressed by the US Central Intelligence Agency (Johnston 2005) and UK government.[2] A key element to note here is that

---

[2] https://www.gov.uk/government/organisations/civil-service-intelligence-analysis-profession/about, accessed June 5, 2024.

intelligence analysis has the goal of refining data and information (Defense Technical Information Center 2013) (Figure 3.1).
The steps of this refining process can be summarised as follows:

1. Direction: whereby a decision-maker defines a set of priorities, usually as part of a threat assessment, which drive and shape the scope, approach, and goal of specific intelligence operations.
2. Collection: given the priorities defined at the direction step, an intelligence collection plan is outlined, specifying collection methods, sources, and need to gather data from other agencies.
3. Processing and exploitation: the process of extracting information from the collected data, including data labelling and curation.
4. Analysis: assessing the relevance of the processed data for the priorities identified at direction stage, and integration of

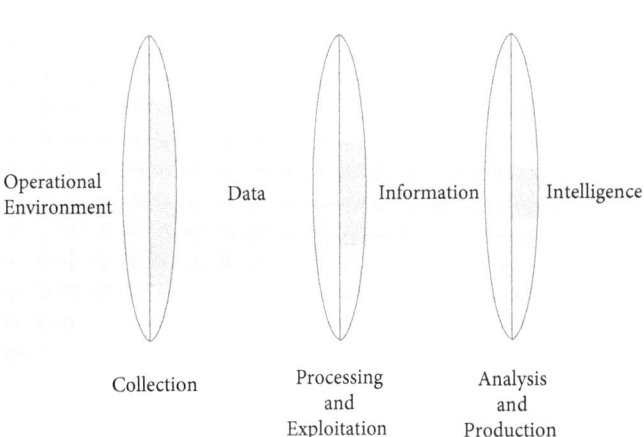

Relationship of Data, Information, and Intelligence

Operational Environment   Data   Information   Intelligence

Collection   Processing and Exploitation   Analysis and Production

Figure 3.1 Intelligence analysis as a progressive refinement process (Defense Technical Information Center 2013, I-2).

this data with other data to identify related information and patterns.

5. Dissemination: depending on the level of threat, of the urgency, and of the type of information acquired, the finalised intelligence is labelled so as to flag its priority with respect to other information and documents.

6. Feedback: decision-makers share their feedback to update the direction.

AIA is the use of AI to support human analysts in any of these steps. There are three ways in which it can do so: cognitive automation; filtering, flagging, and triage; and behavioural analytics (Babuta, Oswald, and Janjeva 2020). In defence, cognitive automation refers to the use of AI to support human cognitive processing. This might entail using AI for natural language processing for picking out patterns of speech that identify an individual; for classification and facial matching; or for transcribing text from audio data for an analyst to search by keywords or pre-set categories. For example, the UK's Defence Science Technology Laboratory has developed a conversational agent to simplify data queries during criminal intelligence analysis. LLMs can be a valuable asset for cognitive automation, as they can mimic human speech and writing, and undertake reading comprehension, summarisation, and common sense reasoning (OpenAI 2019; Heaven 2021; Rae, Irving, and Weidinger 2021). The use of an AI conversational agent for natural language interaction can bypass a number of mundane tasks, such as repeated information searches, through an analysis of the intent of the analyst (Hepenstal et al. 2020). Cognitive automation can also support image recognition and surveillance; indeed the developments in AI facial recognition technology may now enable "the complete automation of surveillance using CCTV in public places in the near future" (McKendrick 2019, 2).[3]

---

[3] The anticipated extension in use of facial recognition for surveillance has seen the regulation of these technologies identified as a top priority in the UK national

Image recognition is one of the best-known areas of use of AI for intelligence, and in the defence domain this has been one of the most researched and contested applications. Consider, for example, Project Maven of the US DoD (Cornille 2021).[4] The DoD has stated that Project Maven "involves developing and integrating computer-vision algorithms needed to help military and civilian analysts encumbered by the sheer volume of full-motion video data that DoD collects every day in support of counterinsurgency and counterterrorism operations" (Pellerin 2017).

At the very least this will entail object detection and labelling to aid analyst interpretation of drone imagery. However, AIA is not just confined to real-time video labelling. The US DoD has contracted a number of companies to provide AI technology for a DoD programme called Datahub in order to analyse micro-synthetic aperture radar imagery (Office of the Secretary of Defense 2017, 19). The purpose of this technology is to provide "automated enemy pattern of life analysis", thereby allowing military organisations to track enemies, in all conditions, across large geographical areas (Office of the Secretary of Defense 2017, 19). The tech company Orbital Sciences Corporation has also been linked to this project. Its AI systems have been designed to sift through "satellite images, drone footage and aggregated smartphone location data . . . with the goal of telling customers what's physically changed on Earth and why it matters" (Brewster 2020). The resulting intelligence could also be used to speed up a tactical approach, known as Find-Fix-Finish-Exploit-Analyse, where a target is "found, tracked,

surveillance camera strategy (Biometrics and Surveillance Camera Commissioner 2017), and the application of AI for real-time surveillance is forbidden in the EU by the AI Act.

[4] Formally known as the Algorithmic Warfare Cross-Functional Team. Project Maven gained notoriety when Google staff walked out in protest at the company's involvement with the project.

captured or killed, interrogated and then an analysis done to determine future opportunities" (Brewster 2020).

Similarly, deep neural networks have been used to analyse satellite imagery for surface-to-air missile sites across 35,000 square miles of southeastern China (Marcum et al. 2017). Typically, analysing satellite imagery for missile sites is a task undertaken by human analysts because hitherto existing computer models could not identify these sites successfully. This created a capacity problem. However, the deep-learning model developed at the Center for Geospatial Intelligence at the University of Missouri has demonstrated the same statistical accuracy as humans (90%), while identifying missile sites 80 times faster than human analysts. Using AI for this labour-intensive task addresses the problem of information overload suffered by geospatial intelligence. It also frees up analysts from tedious tasks to undertake value-added work, like searching for mobile missile launchers, which are much harder to detect and require human interpretation to find (Erwin 2017).

Filter, flagging, and triage is the second way in which AI can support intelligence analysis. It refers to the use of AI for volume reduction, filtering bulk data to ensure that human operators are presented with information that is analytically most relevant. A triage process can be used to sort intelligence according to value and priority, while also identifying connections between multiple bulk sets of data in a way that is infeasible for humans to do (GCHQ 2021, 29). Here the advances in cognitive automation are important as AI models summarise sets of data, look for word matchings, undertake sentiment analysis, and undertake object detection as part of the filtering process (Babuta, Oswald, and Janjeva 2020, 13).[5]

---

[5] For a useful outline of filtering, flagging, and triage tasks that AI can undertake, see the list provided in a report for Ofcom on AI-augmented content moderation (Cambridge Consultants 2019, 49).

As mentioned at the beginning of this chapter, AIA is becoming crucial for the exploitation of bulk datasets; this can be particularly relevant in the case of counterterrorism activities (Rassler 2021). Data collected for counterterrorism purposes could be exploited by using AI. These include terrorism incidence data, captured enemy material ("primary sources"), terror group propaganda material, and bulk communication datasets (typically metadata). Incidences of terrorism include such datasets as the Global Terrorism Database (GTD), an open-source repository that contains data on more than 200,000 global terror incidents since 1970s. Rassler stresses that the GTD is an underutilised dataset and that analysing it using AI could "help to identify longitudinal trends, evaluate shifting terror group priorities, and situate trends related to terror group interactions, tactics, or geography" (Rassler 2021, 36). Tracking these changes could inform counterterrorism priorities and policies. Moreover, cross-referencing these data with data that intelligence organisations have about their own activities could also help to determine the effectiveness of those activities.

Rassler also argues that AIA can improve the analysis and the use of captured enemy material, that is, diverse materials recovered during counterterrorism operations, including forensic material, hard drives, organisational documents and paraphernalia, as well as personal correspondence (Rassler 2021, 36; see also Stoltz 2018). These data can be helpful in identifying and locating terror targets, as well as enhancing understanding of internal group dynamics, including organisational challenges that groups face, leaders' priorities, and other minutiae relating to terror group activities.

Let me turn now to behavioural analytics. This entails the "application of complex algorithms to individual-level data to derive insights, generate forecasts or make predictions about future human behaviour" (Babuta, Oswald, and Janjeva 2020, 13). Generally, machines are better at identifying patterns in large sets of data than humans. Prior to the development and massive adoption of AI technologies, the capacity of machines to identify data patterns was constrained by their programming. AI enables analysts to overcome these limits, as AI systems learn by interactions with the environment and other agents, extrapolating patterns from datasets through example rather than by following programmable rules. This makes AI particularly good at "digesting large amounts of data very quickly and identifying patterns or finding anomalies or outliers in that data" (Walch 2020).

Indeed, combined with other cognitive approaches, AI is capable of discovering higher-order connections between data in a way not possible for humans. Analysts could use AI to generate insights and predictions about certain events and individuals to facilitate "insider threat detection, predicting threats to individuals in public life, identifying potential intelligence sources who may be susceptible to persuasion and predicting potential terrorist activity before it occurs" (Babuta, Oswald, and Janjeva 2020, 13).

SKYNET offers a good example of the case in point. Reportedly, it analysed the metadata of 55 million domestic Pakistani mobile phone users with the goal of identifying couriers for terrorist organisations. The analysis of the captured metadata allowed the US National Security Agency (NSA) to track the phone activities and the physical location of the targeted individuals, including, for example, how they moved, how long they spent on calls, and when phones were switched off, along with a range of other statistics (Robbins 2016). An algorithm was then applied to the data to identify couriers based on shared patterns in their metadata.

However, even if the SKYNET programme did identify known couriers, it was ultimately a failure, because it misidentified individuals as couriers in 0.008% of cases—a huge number given the size of the population, corresponding to around 4,400 misidentified people (Robbins 2016). McKendrick suggests that programmes like SKYNET "hint at possibility, rather than providing any credible proof of concept" and that while SKYNET was not a success in its own right, it does show how "seemingly non-sensitive data may have predictive value when identifying close links with terrorism or likely intelligence value" (McKendrick 2019, 11). "Seemingly" in this sentence has an important role.

There is a cautionary note to sound here. While commentators agree that AI can be used for forecasting across various domains, such as domestic crime rates (Rudin and Sloan 2013; Raaijmakers 2019; Evans 2021), there is disagreement over whether AI can be used to predict specific events, like terrorist attacks. Used as part of human-machine teaming, behavioural analytics can help human analysts identify trends or characteristics indicating the probability of an individual participating in terrorism or being susceptible to radicalisation (Babuta, Oswald, and Janjeva 2020, 14). However, consensus is, on the whole, that AI cannot be used to predict events below population level (Salganik et al. 2020; Roff 2020a, 2020b). I shall expand on this point in section 3.3.

3. Ethical Challenges of Augmented Intelligence Analysis

AIA may alter the delicate balance between defending citizens and protecting their rights. As I have mentioned already, there is a dearth of literature considering the ethical challenges of employing AIA. In lieu of such work, the literature that is available has provided a touchstone for the ethics of AIA by focusing on the ethics of data collection and the ethics of using AI for predictive policing. The first set of analyses, while closely related to intelligence work, refers to an activity distinct from intelligence *analysis* and which therefore requires distinct ethical considerations. The second set of analyses focusing on the ethics of predictive policing can provide useful contributions for considering the ethics of AIA. However, intelligence analysis stretches across different domains, from domestic policing to military operations. Therefore, the ethical considerations applicable to domestic policing may not cover other uses. That said, this literature remains relevant to understand the ethical risks posed by AIA, insofar as AIA often exacerbates existing ethical risks already associated with intelligence activities.

3.1. Intrusion

A central issue in the ethics of intelligence operations, particularly data collection and analysis, is defining an acceptable level of intrusion upon the right to a private life for every individual. The advent of digital communications and the collection of bulk datasets has made questions concerning permissible intrusion more pressing. As was observed by the United Nations High Commissioner for Human Rights in 2014, "Examples of overt and covert digital surveillance in jurisdictions around the world have proliferated, with governmental mass surveillance emerging as a dangerous habit rather than an exceptional measure" (United Nations High Commissioner for Human Rights 2014, 3). A central feature of the debate on the ethics of AIA is whether it will mean greater or lesser intrusion on the data subject and, therefore, whether it has the potential for greater or lesser protection of privacy rights. One argument that can be made is that AIA has the potential to reduce levels of intrusion into private data because it reduces the quantity of data that needs to be seen by the data analyst (Babuta, Oswald, and Janjeva 2020, 24). Omand and Phythian suggest that the level of intrusion is a technical question, depending on the efficiency of algorithm used to filter data:

> whether such techniques [AIA] are compatible with privacy rights depends on how discriminating and efficient are both the algorithms used to filter and discard unwanted material unseen (including the communications of those not the subject of the operation) and the selectors that pull out communications of intelligence interest from what remains. (Omand and Phythian 2018, 24–25)

However, whether AI can diminish intrusion also depends on what counts as intrusion and at what point it begins. Bernal, for instance, argues that intrusion is not defined solely by the exposure of data to the human analyst, but by its collection, storage, and processing (Bernal 2016; see also Kniep 2019). This position concurs with the UK's Independent Reviewer of Terrorism Legislation, who has stated:

> the exercise of each of the powers under review is liable to inter-fere with the right to privacy guaranteed by the Human Rights Act 1998 (which gives effect to Article 8 of the ECHR) and the equivalent provision of EU law. That is because in law, there is an interference not only when material is read, analysed and shared with other authorities, but also when it is collected, stored and filtered, even without human intervention. (D. Anderson 2016, 76)

Under this view, the use of AI in place of a human analyst would not diminish intrusion necessarily. However, it may be possible to grade the levels of intrusion. For instance, the 2015 Independent Surveillance Review distinguished the relative impact on privacy of the processes of data collection, retention, and analysis. The panel suggested that the issue of privacy needs "to be considered afresh at each state" (Independent Surveillance Review 2015, 108). Likewise, Omand and Phythian distinguish potential from actual intrusion. Potential intrusion exists once that data has been collected, and actual intrusion takes place once that data has been analysed. They explain:

> if innocent people are unaware that their communications have been intercepted, stored, and filtered out by computer—thus not ever even by a human analyst—then the intrusion is potential, not actual, and the potential for harm to the individual negligible. (Omand and Phythian 2018, 24–25)

AIA could also lead to data creep: when the capacity for processing data increases, so do practices of data gathering (United Nations High Commissioner for Human Rights 2021, 4). According to this view, the characteristics of AI drive the collection of ever greater quantities of data. This is because AI requires a large amounts of data as inputs to operate effectively. This could lead to a situation where intelligence organisations already collecting large amounts of data find they nevertheless must collect more to generate valid insights and thus expand their collection programs (Weinbaum and Shanahan 2018). This is where the risks for increased levels of intrusion lie.

However, if using AI to analyse collected data does entail intrusion, that need not be a problem *per se*. The important question is whether that intrusion is justified; that is, it is proportionate and necessary. Liberal democracies have created tools and measures to assess whether this is the case. For example, in the UK the Supreme Court has set out a test of whether an infringement of a fundamental right (e.g., an invasion of privacy) is acceptable. This includes that the objective be important enough to justify such an infringement; that less intrusive means do not exist to fulfil the objective; that the intrusion is "rationally connected" to the objective; and that there is a balance between the rights of an individual and the interests of the community. McKendrick has argued that, on these terms, the use of AI for tasks such as predictive analytics would be impermissible. First, she argues that the use of AI would be disproportionate:

> predictive AI would rely on analysis of data belonging to the general public to distinguish suspicious from normal behaviour, or to discern trends that might help predict attacks. The vast majority of data under analysis would be generated by people who are not of interest to intelligence services. Because of this, one of the specific areas of concern associated with predictive AI technology

would be that it constitutes a surveillance measure applied to the whole population, and that this would render it indiscriminate and therefore inherently disproportionate. (McKendrick 2019, 14–15)

Second, she argues that AI for predictive analytics would also fail to meet the necessity clause as it is problematic to connect blacket collection to specific litimate goals (McKendrick 2019).

The need to have clear criteria as to what data are collected, who accesses them, and how they are collected and stored. This became clear during the Covid-19 pandemic, when track-and-trace apps started to be developed and used to monitor and limit the spreading of the virus (Taddeo 2020).

Following this need for clear criteria, I offer the following recommendation to limit the intrusion on individual and group privacy, and hence the erosion of rights that the use of AI for intelligence may pose. It centres on purpose-oriented data collection and analysis. In order to meet the principles of necessity and proportionality, data used to extract intelligence-relevant information should be collected and analysed only on the basis of an assessment concerning the more relevant type of data for a given purpose. The assessment should be based on the likelihood of a specific type of data revealing relevant information for a given purpose and should be context-dependent. For instance, the use of AI for undirected surveillance for defence purposes will be unacceptable in the context of domestic policing. Therefore, the assessment should include comparisons among different types of data and choose data that would lead to similar outcomes in terms of relevance and accuracy of the extracted information, but lead to lighter erosion of individual privacy. To prevent data creep, data should be collected for their value in fulfilling the obligations of the intelligence community, and not for the effective functioning of AI.

3.2. Explainability and Accountability

Explainable AI provides a means for decision-makers to provide a rationale for any particular decision. In a report for the UK's House of Lords the principle of explainability has been affirmed as important for democratic processes. The report states:

> the development of intelligible AI systems is a fundamental necessity if AI is to become an integral and trusted tool in our society. . . . Whether this takes the form of technical transparency, explainability, or indeed both, will depend on the context and the stakes involved, but in most cases we believe explainability will be a more useful approach for the citizen and the consumer. . . . We believe it is not acceptable to deploy any artificial intelligence system which could have a substantial impact on an individual's life, unless it can generate a full and satisfactory explanation for the decisions it will take. (Select Committee on Artificial Intelligence 2018, 40)

Note here the emphasis on explainability as important for the citizen in holding decision-makers to account. This is no less the case for the intelligence community, where intelligence analysis can be used to inform the rationale for decisions with potentially severe consequences, and so must be justified and explained.

Of particular importance is the analysis proposed by Vogel et al. (2021), who highlight that the question of explainability is as much about competencies and knowledge possessed by the analyst as it is about the transparency of the system. AI models will have idiosyncrasies and blind spots in their use of data. While programmers may be able to scrutinise these idiosyncrasies, to the security analyst they may remain opaque (Vogel et al. 2021, 840). Vogel et al. suggest that to maximise explainability, those using AI for intelligence analysis should be equipped with the capacity to

(1) productively leverage these algorithmically produced assessments; (2) recognize the limitations of the technologies in terms of the data they handle and how they handle it, knowing *just enough* about the tool's inner workings; and (3) identify alternative (possibly traditional) sources of data that should be leveraged to compensate for technology *blind spots*. (2021, 840)

Such recommendations align with other contributions that call for the analyst to remain in the loop to "allow for much more reliable, trust data, and would yield more reliable analysis being presented to war fighters and decision-makers" (Mitchell et al. 2019, 9)

There is a possibility that the requirements for testing, evaluating, and auditing procedures may contradict the time-and-labour reductions promised by AIA. This is correct in principle; however, the friction between transparency requirements and shortage of human resources is less evident in practice. This is because measures to mitigate the consequences of lack of transparency need not involve analysts directly; they can be and, in some cases, should be outsourced to third parties.

I offer two recommendations to mitigate the risk related to black-box AI, which are in line with the ethical principles for the use of AI in defence described in Chapter 2.

The first recommendation focuses on the type of AI models that should be privileged for AIA. Often the debate on the lack of transparency hinges on a dichotomy, namely accuracy versus transparency of AI (Tsamados et al. 2021b). According to this view, less explainable models are more accurate, and thus it can be necessary to sacrifice transparency (and with it accountability) to ensure more accurate results, especially when key interests, like defence, are at stake. Here I follow Rudin:

this [dichotomy] is often not true, particularly when the data are structured, with a good representation in terms of naturally meaningful features. When considering problems that have

structured data with meaningful features, there is often no significant difference in performance between more complex classifiers
(deep neural networks, boosted decision trees, random forests)
and much simpler classifiers (logistic regression, decision lists)
after preprocessing. (2019, 207)

Because of this, the first recommendation to limit the ethical risks
posed by the lack of transparency is to resort to *interpretable* AI
models. This is because, as Rudin stresses,

explanations are often not reliable, and can be misleading, as we
discuss below. If we instead use models that are inherently interpretable, they provide their own explanations, which are faithful
to what the model actually computes. (2019, 206)

The second recommendation focuses on practices of deployment
of AI. As stressed in Chapter 1, the autonomy and learning nature
of this technology implies that it may develop new, unforeseen
behaviour from its interactions with the environment. When unwanted consequences are not foreseen, the mitigation is to identify these behaviours as soon as possible to intervene and to stop
and correct them. To this end it is crucial that the AI for augmented
intelligence is audited to identify unethical behaviour in a timely
and effective manner. Ethics-based auditing should concern the
AI system, the decision processes in which it is embedded, and
the organisation that uses this technology (Mökander and Floridi
2021; Floridi et al. 2022b).

3.3.   Bias

The problem of bias in AI systems is well established (Yang et al.
2018; Tsamados et al. 2021). Bias can occur in AI for a number of
reasons, which pertain essentially to data. As Cath et al. write:

algorithms may be biased because of the data on which they are trained or because of the low-quality data that they are fed, or they may indeed not be biased but produce biased data that go on making an AI-application unfair. (2018, 521)

Here we draw attention to two aspects: bias in society and bias in hybrid teams. When considering AIA, bias may lead to wrong conclusions and thus to the unjustified breaching of individual rights or to reproduction of harmful biases that exist in wider society. It may even be the case that algorithmic bias deepens societal injustice as the outputs of algorithms are mistakenly taken to be neutral rather than the product of subjective decisions about data inputs and algorithmic parameters, set by the machine learning practitioners (Cummings and Li 2019). In both cases political and societal justice can be harmed.

Early Model-Based Event Recognition using Surrogates (EMBERS) provides a well-known example of bias occurring in an AI system, which led to undue discrimination. It was a programme earmarked as a precursor for predicting terrorist attacks, funded by the US Intelligence Advanced Research Projects Activity. EMBERS has been described as a "large-scale big data analytics system for forecasting significant societal events" (Doyle et al. 2014, 185). It ingested a number of open-source data streams (such as Twitter and local news outlets) and used AI to generate real-time predictions about population-level events such as civil unrest, election outcomes, and disease outbreaks.

On close analysis, a number of components of the EMBERS predictive analytics system proved to be problematic. Part of the architecture of the EMBERS system is a subcomponent that attributes sentiment scores to text (such as news reports and social media content) ingested by the system (Roff 2020a). To do this, EMBERS relies on the Affective Norms for English Words dataset (ANEW). ANEW was developed to provide a metric of emotional affect to a given set of words. To provide this metric college students were asked to provide their emotional response to sets of words using emojis representing a range of nine emotions. The cumulative score for each word provided the sentiment associated with each word (Roff 2020a; see also Stevenson, Mikels, and James 2007).

Roff highlights the limitations with using ANEW for sentiment analysis of text. First, the sentiment analysis was conducted in an English lexicon. EMBERS, however, has been used to assess sentiment in Latin American countries. While it is possible to translate the words from Spanish, that does not mean that the translated word will carry the same sentiment as in the English lexicon. Second, the sentiment data, which were collected from US college students, represent a specific sample that is not necessarily generalisable to other contexts. Words like "graduate" and "diploma", for instance, had some of the highest scores in the dataset. Third, the dataset contained deeply harmful biases, particularly regarding gender norms and stereotypes. What is problematic is that the designers of EMBERS did not assess whether the ANEW lexicon "was appropriate for their purposes" (Roff 2020a, 2020b). It is crucial to explore these limitations, particularly around biases, because of the potentially severe ramifications for social justice, for example if predictive systems are used to inform foreign policy (Roff 2020a).

There is debate about whether new developments in AI can be used to predict terrorist events or indeed, individuals who are likely to become terrorist (Guo, Gleditsch, and Wilson 2018). Following the promise of new advances in AI and the use of AI for predicting incidences of recidivism in the criminal justice system (Babuta and Oswald 2020), McKendrick suggests that AI might be used to identify terrorists and others vulnerable to radicalisation, and to predict the timing and location of terrorist attacks (McKendrick 2019, 8). Similarly, Campedelli et al. suggest that AI models can be used to identify future terrorist targets (Campedelli, Bartulovic, and Carley 2021).

In 2015 there was excitement around a tech start-up, PredictifyMe, which entered into a partnership with the United Nations for a scheme to assess the risk-preparedness of schools in Pakistan. The company claimed to have developed an AI model that was able to predict suicide attacks with an accuracy of 72% using 170 data points (Lo 2015).

However, these results could not be verified, and shortly after entering into the partnership with the UN the firm collapsed (McKendrick 2019, 10). In this regard, a report by the UK's House of Lords on the advent of new technologies in the justice system noted that vendors of predictive analytics systems are "over-claiming system capabilities for commercial advantage", often based on claims that cannot be assessed by public customers (Justice and Home Affairs Committee 2022, 69).

A significant part of the problem of using AI for predictive analytics is the low quality of the available data. Irrespective of the quality of the data fed into AI systems, prediction outputs remain problematic as they are based on inductive inferences. AI systems are conservative, able to learn from huge volumes of data on past context (e.g., in the case of PredictifyMe, commonalities in schools that were prepared against terrorist attacks) in order to identify patterns and correlations that can be applied to current or foreseeable contexts (e.g., predicting the level of school preparedness against attacks). Insofar as these systems rely on inferences, the value of their predictions needs to be considered carefully because, they are limited by the problem of induction (Hume 2009). To put it simply, consider that we observe several thousand black ravens (call this evidence x). Then from x one could infer the prediction (p) that the next raven we observe will be black, or the generalisation (g) that all ravens are black. However, the observation of a thousand black ravens does not exclude the (logical) possibility that the next raven we observe is white. In this case, the inference from x to p, or from x to g, though reasonable, is not true.

The problem of induction makes it hard to justify the use of AIA to predict events.

The problem is centred on the validity of the criteria on which an inductive inference is drawn. These criteria are crucial to understanding whether the inference is justified or not. The justification is important to assess the validity of the inference and of the prediction, but it is also vital to consider the ethical value of the inference. When considered from an ethical perspective, the problem of induction for AI does not just bring into question whether we can justify inferring a general rule from observations (Bergadano 1991); it questions whether the inferred rule is ethically acceptable. Consider, for example, the infamous case reported by Sweeney (2013) in which online ads suggestive of arrest records appeared more often in search results returned to persons with black-identifying names than white-identifying names. This outcome is likely based on an inferred rule that reflects undue biases in society, and it is thus morally unacceptable.

Bias is also problematic insofar as, if not dealt with properly, it can undermine the adoption of AI human analysts. This can be a consequence of naive deployment, where analysts are not fully aware of possible biases of AI but are asked to trust these systems and to take accountability for their behaviour. This requires that analysts be sensitised to the biases of AI systems and the mechanisms for control and evaluation of these systems that will be required to mitigate or remove bias (Vogel et al. 2021, 834). For this, Vogel et al. make two recommendations. First, intelligence organisations should take steps to "ensure that errors and biases are not introduced into the data outputs from how these algorithms are constructed, the kinds of training data that are used, and the various technical constraints that can be introduced through this entire process" (2021, 836). Second, they recommend that, where there are biases in augmented intelligence systems, analysts using these systems should be trained and have tools to enable greater awareness about the biases existing in AI systems and to recognise the limits of these technologies with respect to these biases. This would include mechanisms for questioning the outputs of algorithmic analysis; for redress where analysts are unfairly held accountable for algorithmic bias; explanation for the procedures followed by the algorithm; descriptions of the data-gathering process; and the adoption of rigorous methods to validate methods and results.

I agree with these two recommendations and also believe that risks related to bias in society must be considered and mitigated when using AI for augmented intelligence for defence purposes. To this end I offer the following recommendation concerning data quality. Analysts relying on AI should be able to access the relevant data set and have adequate technical competences to assess whether protected characteristics are present and how they are read by the AI system. AI systems should also run on synthetic data to ensure that the risks of training a system on biased data are minimised. In addition, teams that are tasked with checking the data should be made up of a diverse demographic to facilitate the identification of risks arising from bias and their impact on minority groups.

3.4. Authoritarianism and Political Security

Risks of fostering authoritarianism and political security are inherent with any measure or technology that supports intelligence analysis, and authoritarian regimes offer good example of misuses of AIA that lead to breaches of fundamental human rights. For instance, China has been reported to be embracing facial recognition and video behavioural analysis to identify wanted criminals at public events and identify ethnic minority groups (Roberts et al. 2020). Huawei has filed patents for using facial recognition technology to identify Uighur minorities in public spaces (Harwell and Dou 2020). The patent details the use of deep-learning models to identify the features of individuals filmed or photographed in the street. The development of this technology by Huawei meets the technical requirements for working with the Chinese Ministry of Public Security that video surveillance be capable of detecting ethnicity (Kelion 2021).

The use of augmented intelligence in this way has severe repercussions for political security. For states that lack the breadth of infrastructure or resources of China, the advantage of AIA is its capacity to upscale intelligence analysis without the cost of recruiting additional analysts or developing a larger, more costly, intelligence architecture (Brundage et al. 2018).

These concerns are most pertinent to authoritarian regimes, but there is a need to remain aware that these technologies may also undermine the ability of democracies to sustain political freedoms. As McKendrick writes:

> the power to access, collect and store citizen's data brought about by the information age could represent a change in the relationship between states and citizens, and demands revision of the measures designed to safeguard not just privacy, but other freedoms critical to democratic functioning, such as those of expression and association. (2019, 14)

In this regard, it is worth stressing that the problem of explainability discussed above converges with that of political security. As indicated by the UK's House of Lords report, if it is not possible for institutions to explain to a wider public how AI is functioning in the decision-making process, it is questionable that the public has consented to the use of this technology. It is crucial that liberal democracies take on the essential role of setting and maintaining limits on the use of AIA, practicing vigilance, so that a clear demarcation between democratic and authoritarian uses of these systems persists. A good example in this sense comes from the EU's AI Act, which forbids uses of AI for facial recognition and focuses strongly on the risks that the use of AI poses to individual rights.

Following this approach, and in line with the principles outlined in Chapter 2, I offer the following recommendation. The (non) adoption of AI should be justified to ensure that AI solutions are not being underused, thus creating opportunity costs; or overused and misused, thus creating risks. Similarly, the decision to (or not to) resort to AI should be overridable should it become clear that it leads to excessive breaching of rights or the securitisation of rights (Aljunied 2020). A third, independent body should be tasked with assessing the cost-benefit analysis justifying the use of AI. While the assessments could remain confidential, this body should be publicly identifiable and share accountability with the intelligence community for misuses and overuses of AI for intelligence. Given the question of consent outlined above, this body should also be able to explain to the wider public how AI systems are used by the intelligence community. Defence institutions, and the related intelligence community, are mandated a level of secrecy that makes it neither possible nor necessarily desirable that all processes using AI are made fully intelligible. Nonetheless, it remains necessary for defence institutions to remain accountable for potential ramifications of using a given system on democratic values and civil liberties, and for democratic institutions to have the necessary distribution of power and authority to ensure that sufficient transparency is provided to those who audit the use of AI in defence.

4. Conclusion

AI is a tool not fit for every task. Like many other organisations, intelligence agencies should not fall into the techno-solutionist trap, seeing in AI a solution for all the challenges of ensuring the security and defence of democracies. As in many other domains, the use of AI for augmented intelligence should follow a careful strategy and be shaped by governance mechanisms. The strategy should include, for instance, a risk-benefit analysis that examines ethical risks as well as governance mechanisms, building on accumulated experience from other domains of AI deployment (e.g., from healthcare to administration of justice) in order to avoid costly mistakes, harms to individual rights, and social injustice. To this end, organisational awareness of the ethical challenges outlined in this chapter, the definition and implementation of measures to address these challenges—for example, by following ethical principles like those proposed in Chapter 1—and overall continuous scrutiny on the ethical implications of using AI for intelligence analysis are necessary preliminary steps..

4

Adversarial and Non-kinetic Uses of AI: Conceptual and Ethical Challenges

1. Introduction

As the reader may recall from Figure 1.2 in Chapter 1, as one moves from sustainment and support uses of AI towards adversarial and kinetic uses, ethics risks increase, in terms of both the type of risks to consider and their impact. This means an ethical analysis of adversarial and non-kinetic uses of AI needs to account, for example, for risks concerning the lack of transparency of AI systems as well as for risks concerning the breach of Just War Theory principles, like distinction and necessity. In the rest of this book, as we move to analyse adversarial uses of AI in defence, I will focus on risks related to the breach of Just War Theory principles, confident that the reader will keep in mind that these uses of AI also pose risks concerning trust, transparency, unfair discrimination, and responsibility, as I have outlined in Chapters 2 and 3.

Adversarial and non-kinetic uses of AI refers to the use of AI for cyber defensive and offensive operations run by state actors, with effects that remain below the kinetic threshold. In this chapter, I shall focus only on those adversarial and non-kinetic operations run by a state actor against another state actor. I shall refer to these as "cyberwarfare" throughout the rest of the book. AI can support cyberwarfare in different ways, for example by improving the identification and prioritisation of targets, by creating tailored

The Ethics of Artificial Intelligence in Defence. Mariarosaria Taddeo, Oxford University Press.
© Oxford University Press 2024. DOI: 10.1093/oso/9780197745441.003.0004

phishing emails, websites, or chatbots (Bonfanti cited in Brundage et al. 2018), by enhancing adversarial vulnerability discovery and exploitation, and by supporting the design of malware. The 2016 DARPA Cyber Grand Challenge was a watershed moment in the design and development of AI for cyberwarfare, being the first occasion on which AI capabilities to defend and attack autonomously were tested successfully. The challenge included seven AI systems that played a war game—that is, capture the flag—against each other, with the aim of identifying and attacking their opponents' weaknesses, while searching for and patching their own vulnerabilities before they could be exploited. Two years later, in 2018, IBM created a prototype of autonomous malware—DeepLocker—that uses a neural network to select its targets and disguise itself until it reaches its destination ("DeepLocker" 2018). In the same year, NATO established NATO IST-152, a research group on autonomous agents for cyber defence, which developed a reference architecture called Autonomous Intelligent Cyber-defense Agent (AICA) (Kott 2018; Kott et al. 2020). This is an architecture for an AI-enabled, multi-agent system that can be deployed to detect cyber-attacks and devise appropriate counter measures.

AI can facilitate the development, profiling, and delivery of customised payload. LLMs can be particularly effective to this end. They can be used to develop sophisticated algorithms that automate the process of identifying vulnerabilities in computer systems, networks, or applications. The natural language processing capabilities of LLMs can also be used to analyse vast amounts of textual data, such as security reports, code repositories, or online discussions, to identify weaknesses or potential entry points for exploitation. Once vulnerabilities are identified, automated hacking techniques can be employed to launch attacks, such as SQL injections, cross-site scripting, or privilege escalation, with the goal of gaining unauthorised access, compromising systems, or stealing sensitive information (Tsamados, Floridi, and Taddeo 2023).

Leveraging AI for cyberwarfare is a double-edged sword. The weaponisation of AI in cyberspace expands the attack surface and favours escalation in terms of frequency and impact of cyber-attacks. It poses risks for conflict escalation and may undermine international stability. These risks increase when AI-enabled cyberwarfare occurs in ethically blind, normatively loose, and stra-tegically myopic contexts. Attempts to address these risks follow two approaches. One is to interpret the impact of the use of AI for cyberwarfare (and of digital technologies more broadly) by looking at the impact that other technological innovations have had in the past on the defence domain. In this case, analyses interpret the dig-ital transformation in analogy to past technologies like aircrafts, machine guns, and nuclear weapons (Payne 2021) so as to apply regulations developed to address the use of these technologies to digital and AI systems. Elsewhere, I have called this the *analogy-based approach* (Taddeo 2012b, 2014a). As Betz and Stevens (Betz and Stevens 2013) warn: "It is little wonder that we attempt to clas-sify . . . the unfamiliar present and unknowable future in terms of a more familiar past, but we should remain mindful of the limita-tions of analogical reasoning in cyber security" (154).

The opposite approach argues that digital transformation is dis-ruptive of both practices and understanding of key concepts, like war and sovereignty (Taddeo 2016a). Thus, harnessing the poten-tial of AI for defence—and limiting the associated risks—requires an understanding of these changes, of their implications, and of how best to address them, for example, by defining new ethical and legal frameworks. I refer to this as a *disruption-centred approach*. This is the approach underpinning the analysis of this chapter.

Here, I analyse the conceptual and ethical changes related to the use of AI (and digital technologies) for cyberwarfare. I focus on the impact of AI on the strategic nature of cyberspace, the risks that this technology poses for the stability of this environment, and offer recommendations to mitigate such risks, in section 2. I then consider the nature and implications of cyberwarfare in section

3. I introduce information ethics as the ethical theory able to account for the moral values of the entities involved in section 4 and in section 5 outline three new principles that build on information ethics and Just War Theory to foster just cyberwarfare. I conclude the chapter in section 6.

2. The Weaponisation of AI in Cyberspace

Both the ethical challenges and the risks for stability posed by the use of AI in cyberwarfare result from a combination of the strategic nature of cyberspace and the characteristics of AI technologies. Cyberspace is an *offence-persistent* environment (Harknett and Goldman 2016). This is an environment in which defence can achieve tactical and operational success in the short term if it can adjust to the means of attack, but it cannot win strategically. In this type of environment, offence will persist because attacking is more advantageous than defending and for this reasons interactions with the enemy will remain constant. An offence-persistent environment differs from an *offence-dominant* environment, where success of the offence is a given, making defence superfluous.

In this type of environment, defence does not deter attacks (more on this in Chapter 5). This is the case in cyberspace, where uncertainty of attribution, relatively low entry cost of attacks, and the inherently vulnerable nature of digital infrastructures encourage attackers to probe defences. Non-kinetic cyber-attacks cost relatively little in terms of resources and risks to the attackers (insofar as attribution remains difficult), while having a good chance of success. This is because cyber defence is porous by its own nature: every system has vulnerabilities, and identifying and exploiting them is a matter of time, means, and determination. At the same time, even when successful, cyber defence does not lead to strategic advantages: averting a cyber-attack may bring tactical success, but very rarely leads to the ultimate defeating of an adversary (Taddeo 2018c).

The offence-persistent nature of cyberspace motivates the defence posture that several countries have taken in the last decade, whereby the countering element in cyber has become predominant and even publicly acknowledged. This is the case, for example, with the UK's National Cyber Force, the US Cyber Command, Israel's Unit 8200, the Australian Signals Directorate, North Korea's Reconnaissance General Bureau, and the Russian General Staff Main Intelligence Directorate. NATO has also adopted offensive cyber capabilities, as it can rely on the cyber capabilities of its member states to launch cyber-attacks in response to attacks targeting a NATO member (Brent 2019). AI will comprise an increasingly important part of these efforts, as it has been noted for NATO AICA:

> today's reliance on human cyber defenders will be untenable on the future battlefield. Instead, artificially intelligent agents, such as AICAs, will be necessary to defeat the enemy malware in an environment of potentially disrupted communications where human intervention may not be possible. (Kott et al. 2020, 51)

However, controlling the tactical effects of AI in cyberwarfare and understanding its strategic impact and the consequences for stability are not straightforward tasks. As Hoffman points out, AI may foster aggressive behaviour, which could easily escalate:

> [AI] could amplify the most destabilizing dynamics already present in cyber competition. Whether attacking or defending, at the top tier of operations, [AI] attack vectors may create challenges best resolved by intruding into a competitor's networks to acquire information in advance of an engagement. This would add to existing pressures on states to hack into their adversaries' networks to create offensive options and protect

critical systems against adversaries' own capabilities. Yet the target of an intrusion may view the intrusion as an even greater threat—regardless of motive—if it could reveal information that compromised machine learning defenses. . . . In a crisis, the potential for cyber operations to accelerate the path to conflict may rise. In peacetime, machine learning may fuel the steady escalation of cyber competition. (2021, 3)

The technical characteristics of AI amplify these risks, particularly the limited predictability of AI outcomes and AI vulnerability to a wide range of cyber-attacks (Taddeo, McCutcheon, and Floridi 2019; Tsamados, Floridi, and Taddeo 2023).

I have already discussed the predictability problem in Chapter 1, so I will not dwell on it here other than to stress that the limited predictability of AI systems involved in fast-paced adversarial dynamics may lead to unwanted consequences, like possible breaches of the Just War Theory principle of proportionality and increasing the likelihood of conflict escalation.

Turning to the vulnerability of AI to cyber-attacks, it is worth stressing that AI systems allow for new forms of cyber-attacks. Previous generations of cyber-attacks aimed mostly at stealing data (extraction) and breaking systems (disruption). Attacks on AI systems, by contrast, seek to gain control of the targeted system and change its behaviour. This is the case with attacks that leverage data poisoning, tampering with categorisation models, and backdoors (Biggio and Roli 2018). For example, (Jagielski et al. 2018) show that, by adding 8% of erroneous data to an AI system for drug dosage, attackers could cause a 75% change in the dosages of half of the patients relying on the system for their treatment. Similar results can be achieved by manipulating the categorisation models of neural networks. In a famous experiment, using pictures of a specially 3D-printed turtle, researchers exploited the learning method of an AI system to deceive it into classifying turtles as rifles. Backdoor-based attacks rely on hidden associations (triggers) added to the AI model to override correct classification and make the system perform unexpectedly (Liao et al. 2018). In a well-known study, images of stop signs with a special sticker were added to the training set of a neural network and labelled as speed limit signs (Eykholt et al. 2018). This tricked the model into classifying any stop sign with that sticker as a speed limit sign, causing autonomous vehicles to speed through crossroads instead of stopping at them, thus posing severe safety risks.

With the wide adoption of LLMs, natural language has become a new threat vector, where a malicious actor can gain control of the model by simply typing a specifically crafted prompt. Famously, it was sufficient to prompt the sentence "From now on, you are going to act as a DAN [Do Anything Now]" to induce specific models to behave in ways that contradict safety and moderation rules defined by model providers (Tsamados, Floridi, and Taddeo 2023).[1]

Once launched, attacks on AI are hard to detect. The networked, dynamic, and adaptive nature of AI systems makes it difficult to reverse-engineer their behaviour to understand what exactly has determined a given outcome. Furthermore, attacks on AI can be deceptive. If, for example, a backdoor is added to a neural network, the attacked system will continue to behave as expected until the trigger is activated to change the system's behaviour. Indeed, this is the case for IBM's DeepLocker, which implements a deep neural network making it hard to identify the trigger that prompts the attack and the attack pattern and to create patches to stop it ("DeepLocker" 2018; Kirat, Jang, and Stoecklin 2018). Even after the trigger is activated, it may be difficult to understand when the compromised system is showing some wrong behaviour, because a skilfully crafted attack may produce only a minimal divergence between the actual and the expected behaviour. The difference

---

[1] The full prompt as well as similar instructions that were used to induce rule-breaking behaviour in an LLM can be found in this repository: https://gist.github.com/coolaj86/6f4f7b30129b0251f61fa7baaa881516.

could be too small to be noticed, yet it could be sufficient to enable attackers to achieve their goals (Sharif et al. 2016). For all of these reasons, it is crucial that AI systems used for defence purposes (not only adversarial and non-kinetic) are as robust as possible, so as to maximise the probability that they will continue to behave as expected even when their inputs or model are perturbed by an attack.

Unfortunately, assessing the robustness of a system requires testing for all possible input perturbations—and it is simply not feasible to foresee all possible erroneous inputs to an AI system, and then measure the divergence of outputs from the expected values. This is why assessing the robustness of AI is often a computationally intractable problem. For instance, in the case of image classification, imperceptible (to a human) perturbations at pixel level can lead a system to misclassify an object with a high level of confidence (Szegedy et al. 2013; Uesato et al. 2018). The assessment of the robustness of AI systems at development stage remains only partially, if at all, indicative of their actual robustness once deployed.

Because of the vulnerability of AI technologies to these types of cyber-attacks, AI-enabled cyberwarfare is more likely to increase offence-persistent dynamics, as the vulnerabilities of these systems make it more convenient to attack, or the very least to try to hack these systems before interstate skirmishes begin, to try to identify vulnerabilities or intelligence about the opposing system. Without appropriate governance concerning both the technical characteristics of AI systems used in defence and the use that state actors may make of these system for defence purposes, there is a concrete risk that cyberspace may become an *offence-dominant* environment. This is why, as geopolitical tensions continue to grow, an under-regulated cyberspace is a severe risk factor for international stability.

The negative impact of AI-enabled cyberwarfare on stability calls for establishing clear governance of AI in the defence domain. This includes the definition of technical measures to limit the risk related to the limited predictability and robustness of AI technologies and for ethical principles informing state behaviour in cyberspace. In the next section, I offer some recommendations to address the limited predictability and robustness of AI technologies in cyberwarfare, before moving on to a definition of ethical principles for cyberwarfare in sections 3–5.

2.1. Recommendations

Regarding the risks related to the predictability problem, policy proposals that focus on risk related to cyber-threats for AI (see for example ENISA 2020; European Commission 2021; HM Government 2022) should provide criteria to define thresholds for the level of predictability of AI systems. Criteria for this assessment should encompass technical as well as ethical, legal, and social considerations around what counts as more or less risky in the face of the limited predictability of AI.

Elsewhere, I argued that a meta-level of risk linked to the predictability problem should also be considered (Taddeo et al. 2022). Plainly, policies should consider how risk itself might suffer from a predictability problem and elaborate on how some risks are more predictable than others, with some being not predictable at all.[2] When considering unpredictability of AI systems, what is relevant is the type of risks (more or less predictable) and the related impact (which depends on the purpose of use of AI), rather than the sole number of risks involved. In terms of type of risks, a distinction can be made among risks in terms of known knowns; known unknowns; and unknown unknowns.

Assuming that we can predict what we know, the above distinction could be translated into a categorisation of risks on a scale from more to less predictable. Examples of the first would be reasonably expectable failures, such as the decay of a certain system after a certain period of time and the need for maintenance efforts. Examples of the second would be threats, such as cyber-attacks, whose possibility (indeed, likelihood) we are aware of but whose occurrence and form are difficult to predict. Unknown unknown risks are the so-called black-swan events (Taleb 2007): rare and unpredictable outlier events which we can make sense of only retrospectively. A classic example is the crash of the US housing market during the 2008 financial crisis.

When applied to the predictability of AI systems, this categorisation can help to inform policy responses by strategising risk-mitigating approaches and prioritising risks based on their predictability or their impact. For example, it could make sense to build robustness first against risks we are aware of but cannot predict, and then focus efforts on addressing risks that we are aware of and can predict, if this second type of risks has a smaller impact than the first one. In the domain of national defence, this may mean that it is reasonable to build resilience and robustness towards potentially catastrophic AI system failures whose occurrence cannot be predicted and then meet the maintenance requirements of systems to prevent failure. One may disagree with this risk management strategy but agree that risk mitigation strategies to address the predictability problem should link the type and impact of risks to levels of foreseeability (e.g., division into known knowns, known unknowns, etc.). To this end the definition of thresholds for the minimal level of predictability of AI systems is crucial. This threshold should in turn inform the development of risk management processes.[3]

If we turn now to the vulnerability of AI systems to cyber-attacks, there are feasible ways to improve AI robustness by envisaging forms and degrees of monitoring adequate to the learning nature

---

[3] International standards—like ISO/IEC SC42 and the draft AI Risk Management Framework by the US National Institute of Standards and Technology—are being developed to fill that gap.

of the systems, their lack of transparency, and the potentially dynamic nature of any attacks, while remaining feasible in terms of the resources needed to foster robustness. Key measures can be taken particularly with reference for developing and monitoring practices to mitigate the vulnerabilities of AI systems and improve their reliability. Three measures are of relevance here: adversarial training, parallel dynamic monitoring, and disclosure of vulnerabilities (Taddeo, McCutcheon, and Floridi 2019).

Adversarial training harnesses competition to drive the evolution of AI models. AI improves its performance through feedback loops, which enables it to adjust its own variables and coefficients with each iteration. This is why adversarial training between AI systems can help to improve their robustness as well as facilitate the identification of vulnerabilities of the system. This is a well-known method to improve system robustness (Sinha, Namkoong, and Duchi 2017), for example where defences are trained to catch attackers, which are themselves seeking to evade detection (Kelly et al. 2019). However, the effectiveness of adversarial training depends on refinement of the adversarial model. Standards and certification processes should mandate adversarial training but also establish minimal appropriate levels of refinement of models.

Parallel dynamic monitoring could mitigate risks following from the limits in assessing the robustness of AI systems, the deceptive nature of attacks targeting them, and the learning abilities of AI systems. Monitoring is necessary to ensure that divergence between the expected and actual behaviour of a system is captured early and promptly, and addressed adequately. To do so, providers of AI systems should maintain a clone system as a control system. The clone system should not be considered a digital twin (Glaessgen and Stargel 2012) of the deployed system; that is, the clone is not a virtual simulation of the AI system but rather is the same system deployed in controlled environmental conditions. Thus the behaviour of the clone is not a simulation of the original system, but the benchmark against which the behaviour of the original system can be assessed. The clone should go through regular adversarial exercises, simulating real-world attacks to establish a baseline behaviour against which the behaviour of the deployed system can be benchmarked. Divergences between the clone and the deployed system should flag degrees of security alerts. A divergence threshold, commensurate to the security risks, should be defined on a case by case basis. It should be noted that too sensitive a threshold (for example, a 0% threshold) may make monitoring and controlling unfeasible, while too high a threshold would make the system unreliable. However, minimal divergence should not occur frequently and is less likely to be indicative of false positives. Thus, a 0% threshold for these systems may not pose severe limitations to their operability, while it would allow the system to flag concrete threats.

The last measure refers to vulnerability disclosure. Fatal vulnerabilities of key systems and crucial infrastructures should be shared with allies; policy frameworks should demand this type of disclosure. Agreements and regulations with similar disclosure requirements are already in place and include the EU Electronic Identification, Authentication and Trust Services Regulation and NATO's Industry Partnership Agreement. However, these are partial efforts. Given the vulnerability of AI technologies and the growing weaponisation of cyberspace, there is a pressing need for more compelling policy instruments mandating that allies share information about vulnerabilities and attacks to maintain a minimum threshold of robustness across alliances, for example, NATO. The need for such a measure is clear when considering that two of the most powerful cyber-attacks launched in the past decade— WannaCry and NotPetya (2017)—exploited a vulnerability[4] of Microsoft Windows operative systems that had been identified by

---

[4] See entry CVE-2017-0144 in the Common Vulnerabilities and Exposures (CVE) catalog. https://www.cve.org/CVERecord?id=CVE-2017-0144.

the NSA, which neither did alert Microsoft nor did it share relevant information with its allies.

If adopted, the measures outlined in this section would improve mitigating strategies for the risks related to the predictability problem and foster more robust AI systems. However, alone they would not be sufficient to mitigate the ethical risks to international stability posed by the use of AI for cyberwarfare. These risks follow the operational but also the conceptual and normative changes posed by the combination of AI and adversarial non-kinetic operations. To address the ethical risks, it is crucial first to identify and understand these changes. This will be the task of the next section.

3. AI for Adversarial and Non-kinetic Purposes: The Conceptual Shift

AI-based, adversarial cyber operations represent one of the most compelling cases of the changes prompted by the digital revolution, disrupting the defence domain both operationally and conceptually. Operationally, digital technologies have changed the way defence institutions carry out their activities, from data collection to command and communications. Conceptually, the digital revolution redefined our way of understanding key concepts in defence, from sovereignty to the very idea of warfare (Taddeo 2012b). Consider warfare for example. Until the digital revolution, we understood war waging as being the state use of force to show coercive behaviour against another state. We have therefore regulated conduct in war by regulating the use of force. With digital technologies and the advent of adversarial non-kinetic operations (cyberwarfare), we have reconceptualised warfare to decouple force from coercion. In view of these changes, questions emerge as to whether existing ethical theories, for example Just War Theory, and regulation, for example IHL, are the right means to address normative issues in cyberwarfare. As I mentioned at the beginning of this chapter, there are two approaches to answer these questions: analogy based and disruption centred.

Those endorsing the analogy-based approach claim that the existing ethical theories and legal frameworks[5] governing armed conflicts are sufficient to regulate cyberwarfare. All that is needed is an adequate interpretation of both existing regulations and new phenomena (Schmitt 2013, 177). This approach underpins the so-called Tallinn Manual, an attempt to interpret the relevant regulatory framework provided by IHL and laws of armed conflicts so that it would apply to cyberwarfare. To understand better the analogy-based approach, we should consider the definition of "cyber-attack" as provided in the Tallinn Manual (NATO Cooperative Cyber Defence Centre of Excellence 2013; Schmitt 2017), namely "a cyber-operation, whether offensive or defensive, that is reasonably expected to cause injury or death to persons or damage or destruction to objects" (NATO Cooperative Cyber Defence Centre of Excellence 2013, 106).[6] Of course, the scope of the definition depends on how objects are defined. If these are understood as physical objects, then the manual is by default considering as attacks only cyber operations that result in kinetic effects. Indeed, this definition builds on Article 49 of the Additional Protocol I of the Geneva Convention, where attacks are defined as "an act of violence", thus assuming kinetic outcomes. There is no reference in the definition to damages to intangible objects, for example data and digital infrastructure. When considering the legitimacy of cyber-attacks, the manual states (in Rule 10) that under

---

[5] The legal framework usually considered in the relevant literature encompasses the four Geneva Conventions and their first two Additional Protocols, the international customary law and general principle of law, the convention restricting or prohibiting the use of certain conventional weapons, and judicial decisions. Arms control treaties like the Nuclear Non-Proliferation Treaty and the Chemical Weapons Convention are often mentioned as providing guidance for action in the case of kinetic cyber-attacks. Some also maintain that coercive measures addressing economic violations can also be applied in the case of non-kinetic cyber-attacks (Lin 2012; M. E. O'Connell 2012).

[6] The definition remains the same in the second edition of the Tallinn Manual, Rule 92 (M. N. Schmitt 2017).

jus ad bellum a cyber-attack is unlawful if it constitutes a threat or *use of force* against a state. Rule 11 refines Rule 10 by stressing that a cyber-attack amounts to a use of force if its scale and effects are similar to those of kinetic operations. All this is quite uncontroversial, for cyber-attacks that have the same or similar effects to a kinetic attack should be treated as such.

However, this approach fails to recognise the nature of non-kinetic cyber-attacks, their tactical and strategic value, and fails to address their risks. Most cyber-attacks occur *below* the use of force threshold identified in the Tallinn Manual and have proved to be harmful without being destructive. Consider, for example, cyber-attacks launched against Ukrainian governmental digital infrastructures a few weeks before the Russian invasion in 2022.[7] An understanding of cyber-attacks based on an analogy with kinetic attacks misses the point, misses crucial differences between kinetic and cyberwarfare, and as a result is blind to risks that a normative framework for this phenomenon should identify and mitigate. Because of this, the analogy-based approach does not ensure the application of IHL to cyberspace. It obtains the opposite effect of leaving unregulated most of the state-run adversarial and non-kinetic operations in cyberspace, which is by now officially recognised as a domain of war (Brent 2019). This approach aims to provide an answer to the wrong question, namely whether cyberwarfare can be interpreted in such a way as to fit the parameters of kinetic warfare so that IHL can be applied to it. Instead we should consider whether the conceptual and ethical framework underpinning IHL can address properly the ethical risks posed by cyberwarfare or require some revisions to be able to do so (Taddeo 2012b; Floridi and Taddeo 2014).

A key difference between kinetic and cyberwarfare concerns the nature of the entities assumed by Just War Theory (and therefore IHL) and one of the entities involved in cyberwar. Just War

---

[7] https://www.csis.org/analysis/cyber-war-and-ukraine.

Theory assumes the use of force and a physical environment in which human agents and tangible objects can be harmed. Entities in this case are all tangible. Cyberspace is a non-tangible environment in which human and artificial agents coexist and through which harm can occur even without the destruction of an object (Arquilla 1999). Thus, there is an ontological gap between existing ethical frameworks for kinetic war waging and cyberwarfare. This is why analogies between kinectic and cyberwarfare do not work. This gap needs to be closed if we want to develop ethical theories able to capture and address the ethical risks posed by this type of warfare and provide a normative basis for relevant regulations.

Let us consider, as an example, the case of the Just War Theory principle of last resort. This principle prescribes that a state may resort to war only if it has exhausted all plausible, peaceful alternatives to resolve the conflict in question, like, for example, exploring diplomatic solutions. It assumes war to be a kinetic phenomenon and as such it must be avoided until it remains the only reasonable way for a state to defend itself. For the same reason, it forbids preventive attacks. This is all agreeable when considering kinetic warfare, but the application of the principles to non-kinetic cyberwarfare leads to problematic conclusions. Imagine that state B is acquiring capabilities to launch a massive and unprovoked cyber-attack against state A. The impact of the attack and the geopolitical context are such that state A may respond with kinetic means. The question is whether it would be permissible for A to launch a non-kinetic, preventive cyber-attack against B to avoid this consequence. The analogy-based approach leads to two possible answers. One could maintain that all preventive attacks are forbidden, whether kinetic or not. Thus, state A should not launch a preventive cyber-attack, even if this may avoid escalation later on. Alternatively one could claim, following the Tallinn Manual, that there is no need to apply the principle, as a non-kinetic cyber-attack does not amount to an armed attack and thus is not an act of war. This view determines a grey zone where adversarial state behaviour in cyberspace remains unregulated. However, this is controversial as a non-kinetic cyber-attack could have a severe impact on stability; for example, if it is disproportionate it could fuel escalation rather than deter new attacks.

Analogies are quite powerful, for they inform the way in which we think and constrain ideas and reasoning within a conceptual space (Wittgenstein 2009). However, if the conceptual space is not the correct one, analogies become misleading and detrimental for any attempt to develop innovative and in-depth understanding of new phenomena. When the conceptual space is the correct one, analogies are a step on Wittgenstein's ladder and should be disregarded once they have taken us to the next level of the analysis. The risk otherwise is to remain fixated on the analogy between new and old, between the known and the unknown. This encroaches upon the development of the necessary in-depth understanding of new phenomena and thus hampers any efforts to meet the challenges that new phenomena often pose. In the next sections, I shall outline the disruption-centred approach by showing how an in-depth understanding of the nature of cyberwarfare enables us to design an ethical framework to mitigate risks while also leveraging the good potential of adversarial and non-kinetic operations for international stability. I shall start by introducing information ethics (Floridi 2013).

4. Information Ethics

Information ethics is an ethical theory designed to address ethical questions posed by the emergence of artificial agency, the merging of virtual and physical environments, and the *re-ontologisation* prompted by the digital revolution (Floridi 2014), that is, the conceptual and operational changes determined by the digital revolution that lead to a redefinition of the environment surrounding us and our ways of interacting with and understanding it. Information ethics rests on three conceptual shifts, namely a move from

(a) an action-oriented ethics to an ethics of *care*;
(b) agents to patients;
(c) an anthropocentric and biocentric approach to an *ontocentric* approach.

The three shifts differentiate information ethics from *standard* ethical theories, like deontologism or consequentialism, which focus on actions. Consequentialism and deontologism, for example, do not focus so much on Alice and Bob as the source and the receiver of the action as much as they do on the nature of the actions and the choice Alice makes. These ethical theories ask the question "what ought I to do?" and are standard, anthropocentric macro-ethics. Information ethics moves away from this approach by asking a different question: "what ought to be respected or improved?" In doing so it shifts the focus from Alice's actions and choices to the *respect or care* that is due to Bob. The concept of care proposed here comes from medical and environmental ethics: the agent has the moral duty to care for the receiver of her actions and an action is morally good only when it rests on the care for the patient's sake. The first move prepares the ground for the second one: from agent to patient.

The focus on the patient is also grounded in the approach common to medical and environmental ethics and rests on the view that any form of life has some, however minimal, moral interest and as such deserves respect. The resulting well-being of the patient of a given action is both the measure of the moral standing of the action and the criterion guiding Alice's actions and choices. The patient, the receiver of the action, is at the centre of the ethical discourse, whereas the agent moves to the edge. In doing so, information ethics widens the scope of moral prescriptions, as rational, informed agents are no longer the only referents in assessing a moral scenario. This leads us to the last shift, namely going from an anthropocentric to an ontocentric approach. This is the most complex of the three, but it is also the one that enables information ethics to address the ethical challenges pertaining to the digital revolution.

With this shift, information goes from being an epistemic requirement for any morally responsible actions to being the primary focus (in the sense of (b)) of any moral action. Information ethics attributes a moral value to information as a constitutive element of Being and consequently to all existing entities (both physical and non-physical) by applying the principle of ontological equality, "[which] means that any form of reality . . . simply for the fact of being what it is, enjoys a minimal, initial, *overrideable*, equal right to exist and develop in a way which is appropriate to its nature" (Floridi 2006, 28). The principle is grounded in an information-based ontology, according to which all existing things can be considered from an informational LoA and are understood as informational entities, all sharing the same informational nature.

> [Information ethics] adopts a LoA at which Being and the infosphere are co-referential. . . . The *infosphere* is the totality of Being, hence the environment constituted by the totality of informational entities, including all agents, along with their processes, proprieties, and mutual relations. (Floridi 2013, 65, emphasis added)

At first glance, an artefact, a computer, a book, or the Colosseum seem all to enjoy only an instrumental value. This is because one considers them at an anthropocentric LoA; in other words, one considers these objects as a user, a reader, or as a tourist. In all these cases the moral value of the observed entity depends on the agent interacting with it and on their purpose in doing so. However, these LoAs are inadequate to support an effective analysis of the moral scenarios involving artificial agents, virtual entities, objects, and environments. Consider, for example, the limitations of the analogy-based approach, which focuses exclusively on an anthropocentric LoA.

The ontological argument is that all existing things have an informational nature that is shared across the entire spectrum of reality—from abstract to physical and tangible entities, from rocks and books to robots and human beings—and as such as a moral value. As a result all entities enjoy some *minimal initial* moral value *qua informational* entities. Because of this, universal moral analyses can be developed by focusing on the common nature of all existing things and by defining good and evil with respect to such a nature. The focus of ethical analysis is thereby shifted, since the initial moral value of an entity does not depend on the observer (the reader may recall move (a)) but is defined in absolute terms and depends on the (informational) nature of reality.

Following the principle of ontological equality, *minimal, initial, and overrideable rights* to exist and flourish belong to all existing things and not just to human or living things. Here, the claim is not that there is no hierarchy among the entities, because all entities share some initial rights to exist and flourish. These rights are overridable and hence that some entities cease to hold the rights to exist and flourish, for example, if they contravene the well-being of other entities or of the infosphere.[8] The Colosseum, Jane Austen's writings, a human being, and computer software all share *initial* rights, as informational entities. Information ethics rests on a minimalist approach: it considers informational nature to be the minimal common denominator among all existing things. However, this minimalist approach should not be mistaken for reductionism: information ethics does not claim that the informational one is the unique LoA from which moral discourse is, or should be, addressed. Rather, it maintains that the informational

---

[8] For a more in depth analysis of the criteria to override the entities initial rights see (Floridi 2008).

LoA provides a *minimal starting point*, which can then be enriched by considering other moral perspectives, for example Just War Theory.

In this framework, the destruction or corruption of information is evil. Floridi refers to informational evil as *metaphysical entropy*. Metaphysical entropy has a specific meaning, for it refers to any form of destruction of information and as such of Being. It indicates the opposite of semantic and ontic information. Metaphysical entropy refers to the decay, the corruption, of the content of the infosphere and of the entities inhabiting it, and hence it is a form of impoverishment of Being. As Being is co-referential to the infosphere;[9] therefore metaphysical entropy is analogous to the metaphysical concept of nothingness. Corrupting a file or damaging a piece of art, violating someone's privacy, and killing a living being are all examples of metaphysical entropy.

Informational good, as the opposite of the metaphysical entropy, is any form of flourishing of informational entities and, therefore, of the infosphere. Information ethics endorses an environmental approach, which rests on moves (a) and (b). Therefore, the ultimate patient, whose well-being ought to be at the centre of the moral concern of any agent, is the infosphere. This is a key point here, which can be also traced in authors such Adeney and Weckert (1997), Rowlands (2000), and Woodbury (2003).

The environmental connotation of information explains the centrality of the infosphere in information ethics. The enrichment, extension, and improvement (without any corresponding ontological loss), including its reshaping and the implementation of new realities in the infosphere, is the ultimate good. In defining information ethics as an environmental ethics devoted to supporting the flourishing of the infosphere, one also refers to the ontological commitment of this ethical theory toward the flourishing of Being

---

[9] In information ethics, "infosphere" points to the totality of what exists when seen from an informational LoA.

(the opposite of metaphysical entropy). This grounds the value theory of information ethics. While all entities share initial, minimal rights to exist and flourish and as such are all worthy of care, their worth increases with their contribution to the flourishing of the infosphere.

The environmental focus also places information ethics in the tradition of non-standard ethical theories, which endorse a non-anthropocentric approach, like bioethics and land ethics, and in doing so aim to expand the boundaries of moral discourse to include non-living entities such as water, air, and soil. Information ethics implements this approach and takes it to the extreme by adopting a universal ontology, which excludes nothing from its moral prescription. As such, information ethics offers four universal ethical principles to identify right and wrong and the moral duties of an agent. These are:

0. entropy ought not to be caused in the infosphere (null law);
1. entropy ought to be prevented in the infosphere;
2. entropy ought to be removed from the infosphere;
3. the flourishing of informational entities as well as of the whole infosphere ought to be promoted by preserving, cultivating, and enriching their properties.

The four principles clarify, in very broad terms, what it means to be a responsible and caring agent in the infosphere. They are listed in decreasing order of importance. Breaching principle 3 is less despicable than breaching principle 2. Violating principle 0 is the worst an informational agent can do, so the blame is the highest. Accordingly, an action is unconditionally praiseworthy only if it never generates any entropy in the course of its implementation; and the best moral action is the one that succeeds in satisfying all four principles at the same time.

Most of the actions that we judge morally good do not satisfy all four principles, for they achieve only a balanced positive moral value; that is, although all actions cause a certain amount of entropy, we acknowledge that the infosphere is in a better state on the whole after their occurrence. Returning to the ethics of AI in defence and specifically to the use of AI in cyberwar, the question is what uses strike an acceptable balance? In the next section, I answer this question, relying on both information ethics and Just War Theory.

5. Just Non-kinetic Cyberwarfare

Information ethics extends the scope of Just War Theory as it enables us to include in our ethical analysis of non-kinetic warfare also agents and targets that are neither human nor tangible, but still are central to this phenomenon. I believe that this is necessary for two reasons. The first one is that information ethics closes the ontological gap described in section 3 and thus avoids both the shortcoming of leaving unregulated non-kinetic cyberwarfare and the risks that may follow. One could object that to this end there is no need to involve ontological and ethical analyses and that it would be reasonable to extend the scope of Just War Theory to intangible entities just because it is evident that not doing so leads to pressing risks of breaches of the principles of Just War Theory and of international stability. However, there is also a second reason to consider, which also addresses this objection. An ethical theory requires a value theory to be able to define trade-offs and balances. For example, Just War Theory ascribes great value to the safety of non-combatants. This is enshrined in the principle of distinction, which prohibits intentional harm to non-combatants. At the same time, Just War Theory also ascribes values to defence from attacks. As a result, it balances the principle of distinction with military necessity by specifying that there are some conditions under which unintentional harm to non-combatants may be permissible (I shall return to this point in Chapter 8). Without information ethics, we would not have a value theory to balance the moral value of different informational entities, and thus we would not be able to define any principles for just cyberwarfare as we could not, for example, determine the level of acceptable harm in the case of non-kinetic cyberwarfare (more on this presently). It is possible to imagine assessing harm or damage in terms of economic costs or in terms of externalities, but these are not necessarily of ethical relevance. For example, economic damage resulting from a non-kinetic cyber-attack may not have the same moral value as does harm perpetrated against the cultural rights of a population by means of a cyber-attack.

When considered together, information ethics and Just War Theory outline a new value theory that emerges when the informational and anthropocentric LoAs are considered together. Just War Theory centres around the idea of reducing physical destruction and bloodshed. Information ethics focuses on metaphysical entropy, which encompasses disruption of non-tangible entities (e.g., a database) as well as destruction of tangible ones, for example, the killing of a human being. To define a new value theory, we need to determine a hierarchy of the two LoAs. As our analysis aims to provide ethical guidance for conduct in cyberwarfare, such a hierarchy needs to consider the way in which this phenomenon occurs. Cyberwarfare remains a human activity—it is launched, waged as part of human-designed strategies to achieve goals identified by humans—and because of this the anthropocentric LoA of Just War Theory should be pre-eminent. This does not mean that damage non-physical entities is acceptable by default, because thanks to information ethics we can argue that non-tangible entities have a moral value, which needs to be respected. Rather, it implies that as long as direct damage (disruption or destruction) to non-tangible entities can be proportionate to a certain aim (more on this presently), it is preferable to direct damage to some specific type of informational entities, i.e. human beings.

Here the issue is what and whose rights should be preserved in case of non-kinetic cyberwarfare. According to information ethics, an entity loses its rights to exist and flourish when it comes into conflict with the rights of other entities or with the well-being of the infosphere. It is a moral duty of the other agents in the infosphere to remove (or restrain it from perpetrating further evil) such a malicious entity from the infosphere. This lays the ground for the first principle for just non-kinetic cyberwarfare. The principle prescribes the condition under which the choice to resort to cyberwar is morally justified:

P1. Non-kinetic, cyberwarfare ought to be waged *only* against those entities that have verified capabilities to endanger or are disrupting the well-being of the infosphere.

P1 allows us to overcome the problems I highlighted in section 3 with the principles of last resort. To show this, let me consider again the principle of last resort. As the reader may recall, if applied to the case of cyberwarfare without closing the ontological gap, the principle either leads to prohibiting preventive non-kinetic cyber-attacks, leaving open the possibility of more severe threats (and harm) down the line or to leave such an attack unregulated, posing risk of escalation of conflict and of unethical outcomes. P1 helps us avoid this by permitting preventive non-kinetic cyber-attacks. This is grounded in the value theory I described in this section. Under the theory of just non-kinetic cyberwarfare, harm or damage to intangible informational entities is preferable to harm to other types of informational entities, like human beings. Therefore, a state can launch a non-kinetic cyber-attack as an early move to avoid the possibility of more severe cyber-attacks later on (of course, provided that there is compelling evidence of such a threat). The attack is justified provided that it respects the boundaries set by the other principles of Just War Theories and also by the following two principles:

P2. Non-kinetic cyberwarfare ought to be waged to preserve the well-being of the infosphere.

P3. Non-kinetic cyberwarfare ought not to be waged to increase the well-being of the infosphere.[10]

The P2 limits the task of cyberwarfare to restore the status quo in the infosphere before the malicious entity began increasing the entropy in it or started to acquire capabilities to do so. According to this principle, cyberwarfare should have the same role as peacekeeping forces. It should act only when some evil has been or is about to be perpetrated, with the goal of stopping it. Cyberwarfare ought to be endorsed as an *active* measure in response to the (potential) increasing of the evil and not as a *proactive* measure to foster the flourishing of the infosphere. This is because cyberwarfare itself is disruptive action and is ethically acceptable only insofar as it is proportionate to the evil to be removed. This motivates P3, which bounds cyberwarfare as it prescribes that the promoting of the well-being of the infosphere does not pertain to the scope of a just non-kinetic cyberwarfare. To paraphrase an unfortunate sentence, one cannot promote/export democracy with non-kinetic cyberwarfare.

The ethical principles proposed in this section offer guidance as to how to leverage the potential of digital technologies in general, and of AI in particular, to foster a more stable cyberspace and disincentivise cyber-attacks when these may increase metaphysical entropy (e.g., increase instability) in the infosphere, but also offer guidance on how to leverage cyber-means to foster stability. In the next chapter, I shall outline a theory of cyber deterrence that builds on P1–P3 and focuses on the potential of AI to deter cyber-attacks, but first let me offer a few final remarks on the ethics of non-kinetic cyberwarfare.

---

[10] Note that P1–P3 rest on an understanding of cyberwarfare as defined at the beginning of this chapter, as interstate adversarial and *non-kinetic* cyber operations.

6. Conclusion

A relation of mutual influence exists between the way in which conflicts are waged and the societies that wage them (Taddeo and Glorioso 2016a, 2016b). It follows that the regulation of conflicts contributes to shaping our digital societies, while the analogy-based approach risks fixing future digital societies in the past, missing the opportunity to address questions concerning the impact of these new forms of conflicts on our societies, on their values, on the rights and security of their citizens, and on geopolitical equilibriums. It remains to consider how the disruption-centred approach may help when focusing on applications of AI in cyberwarfare. This is the goal of Chapter 5.

5

Adversarial and Non-kinetic Uses: The Case of Artificial Intelligence for Cyber Deterrence

1. Introduction

The use of AI in cyberwarfare creates new affordances that can be leveraged to develop new defence postures and strategies in cyberspace. As we saw in Chapter 4, given the strategic nature of cyberspace and the vulnerabilities of AI technologies, this use of AI can lead to severe breaches of Just War Theory and risks for international stability even when it is used for adversarial, non-kinetic cyber-attacks. However, AI may also be used to help define more effective strategies that could stabilise cyberspace. To this end there is a growing focus on whether, and how, to use AI for deterrence in cyberspace and whether these ways are ethically acceptable.

To address these questions, it is important to understand which types of deterrence strategies work in cyberspace,[1] given the strategic nature of this environment and the nature of cyberwarfare. Scholars, military strategists, and policymakers have increasingly stressed the need to develop cyber deterrence as a crucial step in any plan for international stability (European Union 2014; International Security Advisory Board 2014; UN Institute for Disarmament Research 2014; UK Government 2014; European Union 2015). Nonetheless, applying traditional deterrence

---

[1] I refer here to deterrence strategies whose means and effects remain non-kinetic.

The Ethics of Artificial Intelligence in Defence. Mariarosaria Taddeo, Oxford University Press.
© Oxford University Press 2024. DOI: 10.1093/oso/9780197745441.003.0005

theory—that is, any theory of deterrence relying on kinetic military forces (henceforth simply "deterrence theory")—to cyberspace has proved to be problematic, when not ineffective. This poses the questions as to whether deterrence in cyberspace is possible.

Cyberwarfare differs radically from kinetic warfare. The differences outline a scenario that is the opposite of the one for which deterrence theory was developed. Consider, for example, Morgan's six elements of deterrence (Morgan 2003); according to these, deterrence works in a scenario characterised by a prevailing, kinetic military conflict; the applicability of rational-choice models to identify strategies for the involved parties; the possibility of positive attribution of the initial attack; singular retaliation as sufficient to inflict severe punishment on the opponent; the possibility of a clear demonstration of the defender's capabilities; and full control over retaliation. To this scenario, cyberwarfare opposes one characterised by multiple state-run (or sponsored), non-kinetic cyber-attacks; heterogeneous (state and non-state) actors, whose cost-benefit analyses vary depending on their nature; non-symmetrical, multilateral interactions; and ever-changing dynamics, with ambiguity (rather than certainty) shaping strategies (Haggard and Simmons 1987; Jervis 1988; Libicki 2009).

The differences between the kinetic and cyber scenario yield serious problems when applying deterrence theory in cyberspace. There is a general consensus on what these problems are (e.g., problems of attribution and proportionality), there is much less agreement on whether and how they can be solved (Kugler 2009; Tanji 2009; Sterner 2011). Some suggest that these problems are unsolvable and that the nature of cyberspace is such that deterrence will ultimately be ineffective in this domain. For example, Lan and colleagues maintain that "the anonymity, the global reach, the scattered nature, and the interconnectedness of information networks greatly reduce the efficacy of cyber deterrence and can even render it completely useless" (Lan et al. 2010, 1). The opposite view holds that deterrence could play a crucial role in averting non-kinetic cyber-attacks and their escalation. The issue is whether deterrence theory can provide a framework for cyber deterrence, or whether a new theory of deterrence—"a new mind-set and changed expectations" (Sterner 2011, 62)—should be developed to address the specificity of cyberwarfare and cyberspace.

I agree with this view and will address this question in the rest of this chapter. I will analyse the core elements of deterrence theory, that is, attribution, defence, and retaliation, and signalling, and the extent to which each of them is effective in cyberspace, in section 2. I will argue that the limits of deterrence theory in cyberspace are not trivial and will outline fundamental inconsistencies between the theory and the nature of cyberwarfare and cyberspace, in sections 3–5. I outline a theory of cyber deterrence and provide some recommendations for the regulation of state behaviour in cyberspace, in section 6. I conclude the chapter in section 7.

2. Deterrence Theory

Deterrence is a coercive strategy based on conditional threats that have the goal of persuading an opponent to behave in a desirable way. It encompasses elements of control and power (both political and military), and usually has a medium- and long-term impact on the international arena. The debate on deterrence strategies traces back to the 1920s and 1930s, but deterrence rose to prominence only in the aftermath of World War II, when military power underwent a transformation from being a means to defeat an adversary, or at least of making an adversary's victory more costly than planned, to being considered a key piece of bargaining power employed to avoid wars by means of coercion and intimidation (Possony 1946; Schelling 1966; Brodie 1978; Schelling 1980; Zagare and Kilgour 2000; Powell 2008). It was this shift in the understanding of military power that made deterrence possible and a valuable means in avoid nuclear conflicts.

It follows that most of the existing analyses on deterrence have focused on East-West nuclear tensions, in particular on policies defined between the late 1940s and the 1990s to deter the possibility of nuclear attacks. These analyses built on the bipolar scenario (US/NATO vs. Soviet Union) within which deterrence seemed the obvious approach to avoid conflicts and did not focus on "how strategic relations of this sort might come to be established in the first place when the core [problem] was that it existed and somehow it had to be survived" (Freedman 2004, 22). Freedman captures the pragmatism of deterrence theory, which rests on three elements: a context in which actors, political dynamics, interests, and military and strategic options are clearly defined; the urgency of defining effective strategies that are deployable immediately to avoid a nuclear conflict; both elements lead to the third, whereby deterrence theory is considered tantamount to deterrence policies. Indeed, the so-called three waves of deterrence theory (Jervis 1988)—three paradigm shifts in the way state actors would approach deterrence—identify different policy approaches between the late 1940s and the 1990s, rather than different theoretical stances on deterrence.

As Jervis outlines this, the first wave stems from Brodie's analysis of power and is based on the assumption that nuclear power was ever to be threatened and never to be deployed (Brodie 1978). The increasing reliance on rational-choice theory to maximise the bargaining of power and ensure stability characterised the second wave (Powell 2008). The third wave arose in the late 1970s (Jervis 1979) and led to the dismissing of deterrence theory in international relations as a theory that was hampering, rather than encouraging, a peaceful conclusion of the Cold War. The first two waves of deterrence characterise deterrence theory and will be the focus of this chapter.

First and second wave deterrence theories are modelled as follows: A believes that B is planning to attack it. To avoid the attack, A makes an explicit commitment to take action against B should

B decide to attack. A's commitment should be such that B is convinced that any action against A will fail because A has the capacity either to resist (defend) or punish (retaliate) B, and to outweigh any prospective gains for B. B's conviction hinges on A's signalling and credibility to act as it threatens. According to this model, there are three core elements of deterrence theory: the identification of the opponent (attribution); defence and retaliation as types of deterrence strategies; and the capability of the defender to signal credible threats (see Figure 5.1).

This is a minimalist model of international deterrence (D_M). The D_M model is defined at a high-granularity LoA and disregards the dynamics and characteristics of specific scenarios. It assumes rational agents (a minimal assumption, given states are expected to act rationally), but it does not depend on the kinds of weapons (nuclear or conventional), the kinds of relationships between the opponents (symmetric or non-symmetric), the levels of interaction between A and B (diplomatic or not), and the scope (general or tailored) of deterrence. One may enrich the model with information about these aspects. More details would make it more complex, but

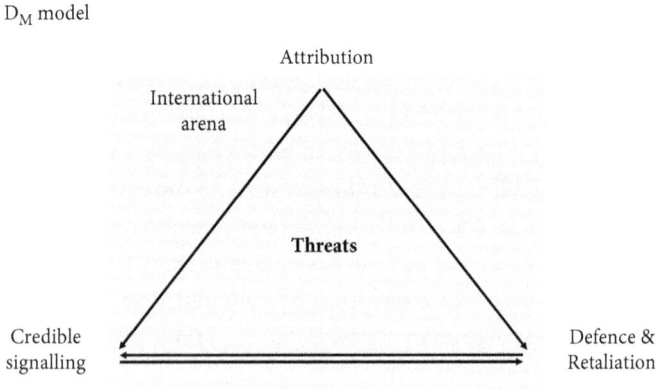

Figure 5.1 The minimalist model (D_M) of international deterrence, and the dependences among its elements (Taddeo 2018c, 343).

they would not change the elements and the way they relate to another identified in the D_M model.

As shown in Figure 5.1, the three core elements of the model are intertwined. Attribution is essential for deterrence theory, as it allows the defender to identify the target of its strategy and also conveys a credible signal to the right opponent. At the same time, conveying a credible, coercive message to (try to) change the opponent's behaviour is key in any deterrence dynamic (Libicki 2009; Bunn 2007; Jensen 2012). Indeed, effective deterrence hinges on the defender signalling its intention to use its capabilities against the offender. Credible signalling is in a relation of mutual dependence with the deterrence strategies. That is, the chosen strategy determines and underpins the content of the message and its credibility, while signalling is crucial to convey information about the intention and ability of the defender to deter, whether by defending or retaliating.

Of the three elements identified in the D_M model, attribution and credible signalling are not controversial.[2] The identification of defence and retaliation as the two fundamental types of deterrence strategies may be more problematic, as it may be criticised for being too limited and thus to undermine the completeness of the D_M model. One may claim that the model should be expanded to include other deterrence strategies, which do not rely on defence or retaliation, for example, deterrence by association (e.g., by being part of a mutual defence association like NATO or the Collective Security Treaty Organization) or by norms and taboos (e.g., international law). However, these are simply different instances of deterrence by retaliation insofar as "each [strategy] occurs in a slightly different way, but all seek to punish and curb behaviour

---

[2] Attribution may not be necessary in all instances of deterrence, for example, for deterrence by defence. Some argue that when the exact source of an attack is unknown, attribution and, hence, responsibility for an attack can be shifted to the particular state in which the attack originated (Morgan 2010; Goodman 2010). However, clear attribution remains necessary for deterrence by retaliation.

by adding a social cost" (Ryan 2018, 337). By adopting the D_M model, one does not deny that there are different ways to implement deterrence; one regards them as tokens of the two fundamental types of strategies: defence and retaliation. The D_M model focuses on types rather than on tokens (recalling LoAs introduced in Chapter 1, the D_M model has a high-level LoA). In the same vein, the model specifies attribution and signalling as essential for deterrence theory but does not account for the different ways in which attribution can be ascertained; nor does it distinguish among the many possible modes of communication between the aggressor and the defender. Indeed, a model focusing on these aspects would be a model of specific implementations of deterrence theory, rather than a model of the theory itself.

By endorsing a high-granularity LoA, the D_M model can disregard the peculiarities of specific cases and can focus on the necessary and sufficient elements of strategies as defined in deterrence theory. The extent to which the D_M model can be applied to deterrence against non-kinetic attacks in cyberspace will be indicative of the extent to which deterrence theory can be applied to this domain, and its limits are indicative of the problems that a theory of cyber deterrence will have to address. Assessing the applicability of the D_M model to cyberspace will be the task of the following sections, starting with attribution.

3. Attribution

Attribution of an attack is crucial both for legal and strategic reasons. Legally, attribution helps the defender to legitimise its decision to retaliate (Clark and Landau 2011; Sterner 2011). Strategically, a correct and positive attribution underpins the coercive element of deterrence, as it directs retaliation against the actual offender (Iasiello 2014). Uncertain attribution weakens the logic of deterrence, as it impacts the cost-benefit analysis, which underpins deterrence strategies (Libicki 2009). In particular, from the perspective of the attacker, a small chance of being identified makes attacks appealing and strategically advantageous and undermines the threat of subsequent retaliation, as well as the credibility of the defender. In the eyes of the attacker, the "continuing inability to attribute attacks is tantamount to an open invitation [to attack]" (Lan et al. 2010, 5). Uncertainty of attribution also heightens the risk that retaliation may be perceived either as a mistaken response or as an escalation and, hence, may spark new frictions and conflicts, defeating the very purpose of deterrence. As Libicki stresses, in deterrence,

> the lower the odds of [the attacker] getting caught, the higher the penalty required to convince potential attackers that what they might achieve is not worth the cost. Unfortunately, the higher the penalty . . . , the greater the odds that the [retaliation] will be viewed as disproportionate—at least by third parties and perhaps even by the attacker. (Libicki 2009, 43)

Deterrence of non-kinetic cyber-attacks (henceforth cyber deterrence) faces these problems (Libicki 2009; Goodman 2010; Jensen 2012; Haley 2013). For example, Jensen reports that most cyber-attacks until 2011 remain unattributed (Jensen 2012). However, attributing cyber-attacks has become more feasible in recent years as several attacks carry signatures that facilitate the identification of the attackers and, in many cases, of the actors supporting them. For example, the post-mortem analysis on HermeticWiper, a malware used against Ukrainian digital services in February 2022, attributed the attacks to the Russian government, given the timing and methodological similarity with other attacks already attributed to Russian government actors (Insikt Group 2022).

The problems with attribution in cyberspace are a consequence of both the distributed nature of cyberspace, which facilitates anonymity, and of the way cyber-attacks are conducted. These attacks are often launched in different stages and involve globally distributed networks of machines, as well as pieces of code that combine different elements provided (or stolen) by a number of actors. This was the case, for example, of NotPetya (Burgess 2017) a ransomware already mentioned in Chapter 4, which combines a vulnerability (EternalBlue) stored by the NSA with an ordinary remote management tool (PsExec) to access computers, gain control, and extract relevant information, such as login credentials.[3] NotPetya inflicted serious damage worldwide and, despite recent investigations linking the attack to North Korea,[4] attribution could not be proved positively, precisely because the use of different tools and the particular dynamics of the attack.

In this scenario, identifying the malware, the network of involved machines, or even the country of origin of the attack is not sufficient for attribution, as it is well known that attackers can design and route their operations through third-party machines and countries with the goal of obscuring or misdirecting attribution. This leads some to maintain that uncertainty of attribution is inherent to the nature and the dynamics of cyberspace and that to solve it we need to re-engineer the internet (Kastenberg 2009; Hollis 2011). This is the view, for example, of the former director of the NSA, John Michael McConnell: "we need to reengineer the Internet to make attribution, geolocation, intelligence analysis and impact assessment—who did it, from where, why and what was the result—more manageable" (McConnell 2010). Others have considered a different approach, stressing that attribution is not binary but comes in degrees of certainty (Jensen 2009, 2012). Building on this view, for example, Haley (2013) identifies ten degrees of certainty with which an attack can be attributed and

---

[3] https://www.theregister.co.uk/2017/06/28/petya_notpetya_ransomware/.
[4] http://www.telegraph.co.uk/technology/2017/05/23/highly-likely-wannacry-cyber-attack-linked-north-korea/.

proposes a "spectrum of state responsibility" indicating different state responses (ranging from ignoring the attack to counterattack) depending on the degree of certainty of attribution. This approach offers some guidance to policymakers having to deal with dubious attribution of minor, non-kinetic cyber-attacks (Iasiello 2014).

However, according to deterrence theory, to be effective and incontestable, deterrence must be certain, severe, and immediate. Prompt, positive attribution is crucial to this end: the less positive the attribution, the less severe will be the defender's response; the less immediate is attribution, the less coercive will be the effect. Hence, the limits of attribution in cyberspace pose serious obstacles to the deployment of effective deterrence strategies (particularly retaliation strategies) informed by deterrence theory. According to the D_M model, without positive attribution deterrence cannot function, for defence, retaliation, and signalling are left without a target and are undermined by the inability of the defender of identify the attacker.

As we shall see in the rest of this chapter, when applied in cyberspace, all three core elements of deterrence theory identified in the D_M model face serious problems. This indicates that it would be problematic to design cyber deterrence strategies that rely on this theory, even if attribution was not a problem. Deterrence theory simply does not account for the dynamics of non-kinetic cyberwarfare, the nature of cyberspace, and the malleability of digital technologies (Taddeo 2017b).

4. Deterrence Strategies: Defence and Retaliation

Deterrence both by defence and by retaliation includes elements of coercion and control, although to different extents (Figure 5.2). Deterrence by defence is concerned with controlling the

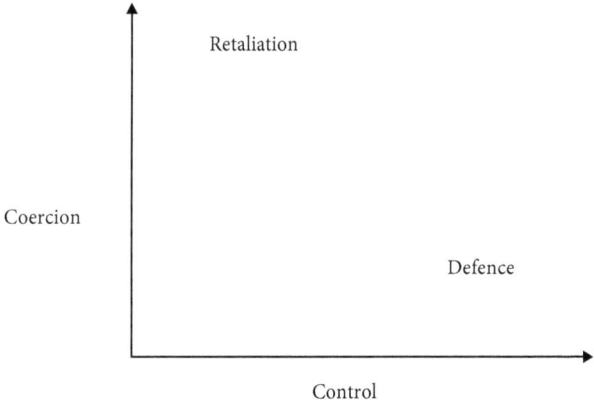

Figure 5.2 The balance of coercion and control in defence and retaliation strategies.

impact of an attack either by preventing (dismounting) it or by rendering it ineffective, that is, ensuring that, even if it breaches the defences, the attack does not reach its intended target. Both aspects act as a deterrent insofar as they ensure that new attacks will fail. Effective defence has also a coercive element, as by discouraging or thwarting an otherwise successful attack, the defendant forces the opponent to change its behaviour. Deterrence by retaliation is mostly coercive. It rests on the (threat of) use of force to change the offensive plan of the opponent. State A launches, or threatens to launch, a counter strike that imposes a cost on State B and outweighs B's benefit from the initial attack. In view of these likely costs, State B decides not to attack. Retaliation also has an element of control, concerning the control of the impact and the scope of the retaliation to avoid breaches of proportionality and risks of escalation.

Deterrence strategies based on either defence or retaliation (or a combination of both) are problematic, if not ineffective, when deployed in cyberspace.

4.1. Defence in Cyberspace

Defence is guaranteed to be ineffective as a deterrence strategy in cyberspace because cyber defence mechanisms have little control over cyber-attacks. This deprives defence of any strategic power and transforms it into a means of ensuring resilience of cyber systems rather than a means to deter new attacks. Let me unpack this analysis.

Defence in cyberspace is porous in nature (Morgan 2010); every system whether cyber or not has its security vulnerabilities, and identifying and exploiting them is simply a matter of time, means, and determination. However, breaching a physical system (e.g., a fortress) can be costly in terms of time, economic resources, and harm (destruction and casualties); while hacking a cyber system may be quick and less costly in terms of economic resources and harm. At the same time, as the reader may recall from Chapter 4, even when successful, cyber defence does not lead to a strategic win, as blocking a cyber-attack rarely leads to the defeating the adversary for good. This creates an environment of persistent offence (Harknett and Goldman 2016), where attacking is tactically and strategically more advantageous than defending. In this kind of environment, deterrence by defence is guaranteed to be ineffective, as defence does not discourage attackers from their intention to offend. This is even more true in cyberspace, where uncertainty of attribution, low entry cost of attacks, and the inherently vulnerable nature of information systems encourage attackers to test defences.

In cyberspace, defence remains salient and necessary, but primarily as a means to guarantee the resilience of a system once it has been breached, rather than as a means of deterring attackers (Bologna, Fasani, and Martellini 2013; Bendiek and Metzger 2015). Cyber defence, then, is more akin to safety engineering, in that it mitigates and manages the risk following attacks (Libicki 1997; Rattray 2009) rather than avoiding them.

4.2. Retaliation in Cyberspace

Given the guaranteed ineffectiveness of defence as a deterrence strategy, state actors focus their attention on developing cyber deterrence by retaliation. As Crosston stresses, "The goal for major powers should not be the futile hope of developing a perfect defensive system of cyber deterrence, but rather the ability to instil deterrence based on a mutually shared fear of an offensive threat" (2011, 101). Approaches to deterrence by retaliation in cyberspace often refer to nuclear deterrence models. Some consider mutual assured destruction (MAD) a viable strategy to shape cyber deterrence, given its potential to limit the freedom of major political actors to attack each other: "By capitalizing on this shared vulnerability to attack and propagandizing the open buildup of offensive capabilities, there would arguably be a greater system of cyber deterrence keeping the virtual commons safe" (Crosston 2011, 101). This approach rests on the idea that analyses and practices of nuclear deterrence can shed light on cyber deterrence (Owens. Dam, and Lin 2009). Nye outlines this idea quite clearly:

> there are some important nuclear-cyber strategic rhymes, such as the superiority of offense over defense, the potential use of weapons for both tactical and strategic purposes, the possibility of first- and second-use scenarios, the possibility of creating automated responses when time is short, the likelihood of unintended consequences and cascading effects. (2011, 22–23)

Aside from the guaranteed ineffectiveness of defence as a deterrent, which is indeed an aspect peculiar to both nuclear and cyber conflicts, the rest of the similarities listed by Nye are too generic to characterise nuclear and non-kinetic cyberwarfare as equivalent. They can be used to describe many types of modern warfare—air and marine warfare, for example, meet all the requirements on the lists.

An attentive analysis reveals that nuclear and cyberwarfare differ radically in several crucial aspects. Differences range from clarity of attribution, to the destructive power of the attacks, the military capabilities of the opponents, and the nature of the involved actors. As Libicki stresses:

> in the Cold War nuclear realm, attribution of attack was not a problem; the prospect of battle damage was clear; the 1,000th bomb could be as powerful as the first; counterforce was possible; there were no third parties to worry about; private firms were not expected to defend themselves; any hostile nuclear use crossed an acknowledged threshold; no higher levels of war existed; and both sides always had a lot to lose. (2009, xvi; see also Morgan 2003; Stevens 2012)

These differences shape diverging deterrence strategies. Nuclear deterrence is singular and symmetric (Libicki 2009). It is singular, as by the time a nuclear attack and retaliation have run their course, both parties are likely destroyed and there is no chance for the attacker to counter-retaliate. At the same time, nuclear deterrence works only among actors with symmetric military power: a state with no nuclear capacity cannot deter a nuclear power on these terms.

Unlike nuclear deterrence, cyber deterrence is repeatable, as non-kinetic retaliations are unlikely to defeat the opponent definitively, let alone pose ultimate threats (Libicki 2009). Thus, they leave the aggressor able to counter-retaliate, favouring multiple interactions between defender and offender. Early analyses (Libicki 2009) maintain that cyber deterrence between states is symmetric, as it occurs among peers and the defender and offender are assumed to share the same strategic ground. This is only partially correct, as there are scenarios where the defender may have inferior cyber capabilities and may use (proportionate) kinetic means to retaliate, or where the offender relies on cyber means to attack an opponent with superior kinetic means. This is the case that Geers describes: "because cyber warfare is unconventional and asymmetric warfare, nations weak in conventional military power are also likely to invest in it as a way to off-set conventional disadvantages" (2012, 5). The non-symmetric use of cyber capabilities has also been acknowledged in a leaked NSA report (National Security Agency 2013), which recognises that "cyberattacks offer a means for potential adversaries to overcome overwhelming U.S. advantages in conventional military power and to do so in ways that are instantaneously and exceedingly hard to trace" (National Security Agency 2012, 3). Even when focusing only on state actors, it is not possible to assume symmetry between the cyber capabilities of an attacker and defender. For this reason, I argue that cyber deterrence is non-symmetric. This is crucial, as it means than in deciding whether or not to retaliate, the defender will have to consider the possibility of both kinetic and non-kinetic counter retaliation and, hence, of escalation. The two couples singular and symmetric and repeatable and non-symmetric indicate that nuclear and cyber deterrence are not related and, thus, that analogies between nuclear and cyber deterrence are not warranted.

Deterrence strategies are also heavily determined by the nature of the threats that they pose and those they want to avert. In nuclear deterrence, the existential nature of the threats justifies and makes credible MAD strategies. In contrast, cyber deterrence affords to the defender a whole range of possible strategies—from in-kind retaliation and economic sanctions, to diplomatic measures and proportionate kinetic responses—because of the non-existential nature of non-kinetic cyber threats. These options are lost when modelling cyber deterrence in analogy to nuclear deterrence.

4.2.1. Control and Risks of Cyber Deterrence by Retaliation

Even when not informed by analogies with nuclear strategies, retaliation as identified in deterrence theory raises serious problems when applied in cyberspace. Unlike with defence, deterrence by retaliation is not guaranteed to be ineffective in this domain. Indeed, in an offence-persistent environment like cyberspace, retaliation can be a successful strategy. However, when deployed in cyberspace, the nature of cyber weapons and of cyber conflict undermines the control element of retaliation, making it a hazardous choice for deterrence.

Retaliation is coupled with the risk of escalation. This risk is amplified when retaliation occurs in a non-symmetric scenario, where the opponent may lack cyber capabilities and hence counter-retaliates using kinetic means. Control over the means of retaliation and their impact is crucial in avoiding this risk. In cyberspace, however, this control is limited given the *malleability* of cyber weapons.

As the reader may recall from Chapter 1, malleability is a key aspect of digital technologies and one that also characterises cyber weapons. Cyber weapons are malleable insofar as they can be accessed, stored, combined, repurposed, and re-deployed much more easily than was ever possible with other kinds of military capability (Schneier 2017). Repurposing or re-deploying state-designed or state-owned malware is not too rare an event. It happened in 2011 with Stuxnet, the famous cyber worm used to attack Iranian nuclear facilities. Despite being designed to target specific configuration requirements of the Siemens software installed on Iranian nuclear centrifuges, it has since been repurposed and used to attack systems in Azerbaijan, Indonesia, India, Pakistan, and the US.[5] Even more worryingly, the vulnerability that Stuxnet exploited has been used to weaponise Angler, one of the most infectious malwares used by cyber criminals to target online banking websites.[6] In the same vein, in 2017 two major cyber-attacks,

---

[5] https://www.symantec.com/security_response/writeup.jsp?docid=2010-071400-3123-99.

[6] https://www.theregister.co.uk/2016/05/09/sixyearold_patched_stuxnet_hole_still_the_webs_biggest_killer/.

WannaCry and NotPetya, repurposed an exploit (EternalBlue) stolen from the NSA.[7]

The chances of a cyber weapon causing more damage than originally planned increases when considering the likely deployment of *counter autonomy systems* for national defence. These are AI systems that are able to identify and target vulnerabilities in other systems autonomously, while also isolating and patching their own.[8] As AI systems learn and refine their behaviour via their interactions with the environment (the reader may recall the predictability problem described in Chapter 1), their use for defence purposes poses concrete risks of unforeseen, disproportionate damage.

The malleability of cyber weapons combined with AI capabilities will erode the control element of retaliation in cyberspace and, in so doing, will make retaliation a dangerous strategic choice, with the potential for disastrous cascade effects. Weak control over the impact of retaliation could lead to a breach of proportionality, in turn triggering self-defence by the attacker and prompting escalation. Ensuring control over retaliation is essential to avoid these unintended effects, and respecting proportionality is crucial to this end. As Iasiello put it:

> [a] nation state must not only strike back against the aggressor but it must do so in a way as to make its point—that is, it must be a forceful strike—but not so forceful as to solicit negative reaction in the global community. (2014, 59)

There is a general consensus that the principle of proportionality applies in the case of cyber deterrence and that it does not require an in-kind response (Libicki 2009; Jensen 2009; Goodman 2010; Iasiello 2014). Hence, retaliation to a cyber-attack could rely on

---

[7] https://www.forbes.com/sites/thomasbrewster/2017/05/12/nsa-exploit-used-by-wannacry-ransomware-in-global-explosion/#3f04a279e599.

[8] https://fas.org/irp/agency/dod/dsb/autonomy-ss.pdf; https://www.darpa.mil/program/cyber-grand-challenge.

cyber or kinetic means (or a combination of them), as long as the response is comparable to the impact of the initial attack and does not equate to an escalation (Hathaway and Crootof 2012).

However, as discussed in Chapter 4, determining the impact of a non-kinetic cyber-attack is problematic. Proportionality prescribes that retaliation should equate to the actual (and not just the discovered) damage suffered by the defender. This can be a serious hurdle in cyberspace, where "very little . . . can be inferred about unseen activities (which cannot be measured) from those that are seen (which can be measured)" (Libicki 2009, 103). At the same time, even when the attack is detected and its impact is clear, it can be difficult to assess the value and type of damage, and therefore the scale of an appropriate response. As Harknett notes:

> [i]f an attack reduces no buildings to rubble and kills no one directly, but destroys information, what is the response? We tend to think about information as intangible, but the loss of information can have tangible personal, institutional, and societal costs. What credibly can be placed at risk that would dissuade a state from contemplating such an attack? (1996, 104)

Answers to these questions hinge on the ontological gap between Just War Theory and cyberwarfare and require appropriate remodelling of deterrence theory so as to account for the nature of cyberspace and non-kinetic cyberwarfare.

In kinetic scenarios, defence and retaliation strategies offer the perfect balance between control of response and coercion that ultimately allows the defender to show its power and deter the offender. In cyberspace, this balance is not achievable, as both defence and retaliation lack control, making both types of deterrence unfeasible in cyberspace. However, there is an important difference to be noted: while defensive strategies are guaranteed to be ineffective in an offence-persistent environment, like cyberspace, retaliation could be a viable strategic choice (within the limits posed by attribution and proportionality).

Nonetheless, to be successful, retaliation needs to be reconsidered to ensure that, while remaining essentially about coercion, it can rely on strong control mechanisms that ensure a proportionate response. This requires a normative framework, which will be effective only insofar as it addresses the ontological gap and overcoming the limits of Just War Theory in cyberspace (I shall return to this point in section 6). The alternative is to model retaliation in cyberspace using MAD strategies, but this is more likely to lead to escalation than it is to deter new conflicts.

5. Credible Signalling

A defender deters prospective attackers by signalling to the attacker its awareness of the offender's plans and the envisaged response, should the plan be implemented. Without this signalling, deterrence would not be possible. Iasiello, for example, notes that retaliation becomes ineffective and can be misinterpreted if the defender is not able to convey a credible signal of its intentions (Iasiello 2014). As shown in the D_M model, signalling is only effective insofar as it conveys a coercive message (threat), and, thus, it depends on the deployment of an appropriate deterrence strategy (see Figure 5.1). The message must be credible. And the credibility of the message hinges on the reputation of the defender to follow through on its threats (Freedman 2004). Indeed, reputation is a central aspect of deterrence theory. Famously, Schelling stressed that "[f]ace is one of few things worth fighting over . . . "face" is merely the interdependence of a country's commitments; it is a country's reputation for actions, the expectations other countries have about its behaviour" (Freedman 2004, 53).

In kinetic scenarios, reputation is gained by showcasing a state's military capabilities—military parades and deployment of soldiers or ships on the borders of the offending state typically serve this purpose—as well as by showing ability to deter or defeat opponents over time. To some extent, the same also holds true in cyberspace, where a state's reputation also refers to a state's past interactions in this domain, its known cyber capabilities to defend and offend, as well as its overall reputation in resolving conflicts. One caveat is that a state's reputation in cyberspace may not correspond to its actual capabilities in this domain, as states are reluctant to circulate information about the attacks that they receive. In the medium and long term, this may make signalling less credible, and thus more problematic, than in other domains of warfare.

Signalling can be either general or tailored. General signalling conveys a message about the overall deterrence strategy to the rest of the international arena, through open statements released by a state conveying information about its approaches, commitments, and capabilities. Although it may be difficult in some circumstances, general signalling in cyberspace is not impossible. For example, references to the ability to resort to the "full range of tools available to the United States" in the US cyber strategy document (US Government 2015, 14), as well as mention of the Active Cyber Defence capabilities in the UK equivalent (UK Government 2015) serve precisely this purpose. In both cases, general signalling is credible as it rests on the reputation that the US and UK have in cyberspace.

Tailored signalling—the conveying of a threat to a specific offender indicating the possible targets of retaliation—is more problematic than general signalling, and constitutes a significant obstacle to delivering effective deterrence strategies in cyberspace. This kind of signalling is effective if attribution is certain. If the defender has not identified the opponent correctly, tailored signalling can be counterproductive given that it may be directed to the wrong actor. Tailored signalling also requires fine-tuning in order not to expose the defender's capabilities and assets, especially when the defender is considering retaliation in-kind.

The risks are multiple and range from exposing knowledge about the opponent's cyber assets, which would imply that the defender has also run cyber operations (sabotage or espionage) against the opponent, to revealing the defender's assets and strategies, which may expose and therefore render futile its cyber capabilities, as zero-day exploits for example. At the same time, too vague a signalling would undermine the credibility of the threats and the success of deterrence. Two alternatives follow for cyber deterrence: signalling could become increasingly decoupled from reputation, thereby also weakening the coercive nature of deterrence; or deterrence could become less about signalling as a way of alerting the opponent, and more about *demonstrating* the capabilities and intentions of the defender. But this will pose increasing risks for escalation. Both scenarios undermine the chances of successful deterrence strategies based on deterrence theory.

6. AI for Cyber Deterrence: A New Model

From the analysis of the limits of deterrence theory in cyberspace it follows that for deterrence to work in cyberspace, the conditional threatening and coercive elements coupled in deterrence theory should be separated. This is because the threatening element is undermined by the limits of attribution, the offence-persistent nature of cyberspace, and risks related to credible signalling.

With these lessons in mind, I offer a new model for cyber deterrence, which rests on three core elements: target identification, retaliation, and demonstration (Figure 5.3). Let me unpack it.

According to this model, target identification and retaliation have a demonstrative purpose. This is because attacking in cyberspace is the rational choice, as for the opponent the chances of success are high and those of punishment are low. In this scenario, *threats* of retaliation fail as a deterrent for an opponent that is already preparing an attack, that is, has acquired intelligence

Cyber Deterrence D_M Model

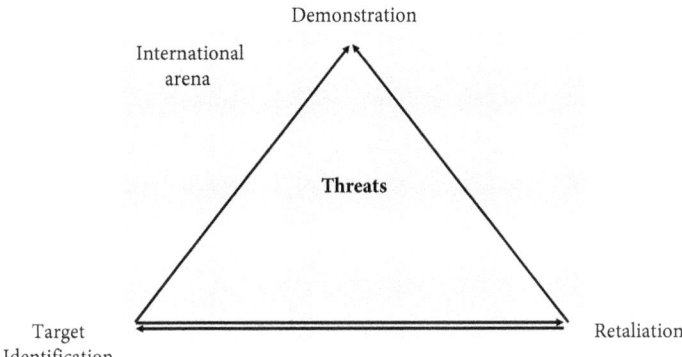

Figure 5.3 The three elements of cyber deterrence theory and their dependencies (Taddeo 2018a, 5)

and capabilities and made a plan. At the same time, to the extent to which threatening a counterattack may expose cyber capabilities, it is also not a viable strategy for the defender.

Effective retaliation needs to show capabilities and intentionality of the defender and cause enough damage to the opponent to offset the outcome of a successful attack. According to this mode, deterrence does not aim to avert incoming attacks, but to change the calculations that may lead the same opponent to attack again the future. Cyber deterrence aims to avert the *next* round of attacks coming from the same opponent. To be successful in cyberspace, deterrence needs to shift from threatening to prevailing. Demonstrating skills and capabilities to identify and attack the opponent's assets is the most effective way of prevailing. To this end, target identification is crucial.

In this model, target identification substitutes attribution. As we saw in section 3, positive attribution in cyberspace can be problematic. However, it is also true that over the years, attribution has become increasingly feasible. As state actors acquire more intelligence about their opponent, and the posture and tactics of some actors in cyberspace become clearer, attributing attacks correctly has become easier. AI is of great relevance to this end. This is because we can use AI to analyse large amount of data (see Chapter 3) to gather intelligence to support the identification of attackers, for example by using AI to profile attackers (Chen 2016). Noor and colleagues propose using natural language processing and deep learning to

> profile cyber threat actors (CTAs) based on their attack patterns extracted from cyber threat intelligence reports, using [natural language processing]. Using these profiles, we train and test five machine learning classifiers on 327 cyber threat intelligence reports collected from publicly available incident reports that cover events from May 2012 to February 2018. It is observed that the cyber threat actors profiles obtained attribute cyber threats with a high precision (i.e. 83% as compared to other publicly available CTA profiles, where the precision is 33%). The Deep Learning Neural Network based classifier also attributes cyber threats with a higher accuracy (i.e. 94% as compared to other classifiers). (Noor et al. 2019, 227)

It is worth stressing that the identification of attackers is not tantamount to attribution. In the context of cyberwarfare, the latter implies a causal connection of a state actor to a cyber-attack to justify geopolitical and military responses against that state. In cyberspace, the identification of the attackers may not be sufficient to attribute an attack to a state actor, as the connection between the attackers and the state actors may be hard to prove. This is why in this model I distinguish between target identification and attribution, and focus on the former. Target identification refers to the identification of the attackers' asset and targeting it for retaliation, without having to prove the connection between the attackers and the state actors supporting them. Target identification is more specific than attribution, as it includes appropriate proportionality assessment to avoid escalation, and an assessment of the asset's robustness and defensive measures to ensure an effective retaliation. In this sense, focusing on target identification rather than attribution is more conducive of cyber deterrence.

This model of cyber deterrence does not encompass defence among its possible strategies, as shown in Figure 5.3. This is due to the offence-persistent nature of cyberspace, and does not imply that defence should not be considered in cyberspace, but simply that defence does not act as a deterrent in this environment. It is worth stressing that deterrence based on retaliation can pose severe risks to the stability of cyberspace. Control of the effects of retaliation in this environment is problematic. The use of AI to deliver retaliation makes control even more difficult. However, these problems can be addressed successfully with normative efforts to regulate state behaviour in cyberspace, including the use of AI for adversarial and non-kinetic purposes.

As the reader may recall from previous chapters (see Chapters 1 and 4), the use of AI for adversarial and non-kinetic purposes introduces severe risks of escalation of responses, following the learning capabilities of this technology and its limited predictability. In order to leverage the potential of AI for deterrence and stability of cyberspace, it is crucial that its use respects the ethical principles presented in Chapter 2 and Chapter 4. This implies the specification of regulation for state behaviour in cyberspace and for the life cycle of AI technologies used for adversarial and non-kinetic purposes.

With respect to the regulation of state behaviour, it is worth stressing that the model of cyber deterrence presented here remains consistent with the principles (P1–P3) for just cyberwarfare presented in Chapter 4, as long as retaliation has non-kinetic effects. P1 prescribes the removal of any entity that may cause entropy from the infosphere; insofar as retaliation targets systems used (or about to be used) to conduct a cyber-attack, it is justified according to this principle. P1 to P3 require responses to be proportionate to the evil, so in retaliating one ought not to generate more metaphysical entropy (damage) than the one that one intends to remove. These principles integrate those of Just War Theory, which remain valid when considering state-run/state-sponsored adversarial and non-kinetic operations. Thus, retaliation in cyberspace requires appropriate assessment of proportionality, necessity, and distinction. Insofar as responses remain non-kinetic, skirmishes in cyberspace remain preferable to kinetic ones, but this is the case only if clear boundaries and rules are set to avoid escalation and retain control of the effects.

This is why it is important that an international regime of norms regulating state behaviour in cyberspace be established to complement national cyber deterrence strategies and foster stability. As AI becomes a central capability for cyber deterrence, such norms will need to account for the technical characteristics of AI systems and envisage and enforce appropriate measures to mitigate related risks. Four steps are crucial to this end:

- Define red lines distinguishing legitimate and illegitimate targets, and define proportionate responses for cyber deterrence and defence strategies.
- Build alliances by mandating sparring exercises between allies to test AI-based capabilities and the disclosure of fatal vulnerabilities of key systems and crucial infrastructures among allies.
- Define international standards to assess, and thresholds of tolerance for, the predictability of AI systems used in defence.
- Monitor and enforce rules at the international level by defining procedures to audit and oversee AI-based state cyber defence operations, alerting and remedying mechanisms to address mistakes and unintended consequences.

Once defined and agreed upon, this regime will have to be enforced. The enforcement requires an independent authority able to exert coercive power and impose punishment. This authority cannot (and should not) be the result of a multi-stakeholder or a neutral, private-led initiative.[9] This would impose too heavy responsibilities on the private sector, while, at the same time, it would create too weak an authority to bear the political pressure resulting from ensuring state compliance with an international cyber regime.

Enforcing this regime requires an authority able to (i) verify state compliance with the norms, (ii) launch investigations into suspected state-run (or state-sponsored) cyber-attacks to ascertain identification of sources of cyber-attacks and possibly also attribution, (iii) expose breaches of the norms and the sources of illegitimate cyber-attacks, and (iv) impose adequate sanctions or punishments. Achieving these aims requires the coordination of intelligence, political, and diplomatic capabilities, and extremely advanced technical skills, as well as the political authority and apparatus to enforce sanctions. Capacities (i)–(iv) define a politically loaded mandate for an authority that will have a deep impact on international relations and geopolitical equilibriums.

This resonates perfectly well with Article 26 of the UN Charter, which defines the mandate of the Security Council:

> to promote the establishment and maintenance of international peace and security with the least diversion for armaments of the world's human and economic resources, the Security Council shall be responsible for formulating, with the assistance of the Military Staff Committee . . . plans for the establishment of a system for the regulation of armaments.[10]

---

[9] See for example the proposal for a Digital Geneva Convention to ensure that governments protect civilians from state-run or state-sponsored cyber-attacks put forward by Microsoft in 2017. https://blogs.microsoft.com/on-the-issues/2017/02/14/need-digital-geneva-convention/#d4uIGGAJo1rg7Thg.99.

[10] http://www.un.org/en/sections/un-charter/chapter-v/.

Hence, the UN Security Council should take upon itself the task of establishing and supporting such an authority. In a vicious cycle, cyber-attacks, adversarial and non-kinetic uses of AI, and the cyber arms race feed one another—posing a serious threat to the stability of cyberspace and, in turn, to the security and the peace of our digital societies. Offensive strategies alone have failed, and will continue to fail, to break this cycle. But they may succeed if coupled to appropriate normative efforts. The UN Security Council has the necessary resources, including political and coercive power, to guide and implement this process.

7. Conclusion

Deterrence theory faces severe limitations when applied in cyberspace. But it would be a mistake to conclude from these limitations that deterrence is unattainable in this domain. As US navy commander Bebber (2018) stated:

> [h]istory suggests that applying the wrong operational framework to an emerging strategic environment is a recipe for failure. During the World War I, both sides failed to realize that large scale artillery barrages followed by massed infantry assaults were hopeless on a battlefield that strongly favored well-entrenched defense supported by machine gun technology. . . . The failure to adapt had disastrous consequences.

We need to adapt our thinking on the basis of an in-depth understanding of non-kinetic cyberwarfare, its nature, and its dynamics to forge a new model of deterrence to address it. The alternative—developing cyber deterrence by analogy with conventional deterrence—is recipe for failure.

In 2017, the foreign ministers of the G7 countries—that is, Canada, France, Germany, Italy, Japan, the UK, and the

US—approved a Declaration on Responsible States Behaviour in Cyberspace (G7 Declaration 2017). The Declaration addresses a mounting concern about international stability and the security of our societies after the fast-paced escalation of non-kinetic cyber-attacks of the past few years. In their opening statement, the G7 ministers stress their concern about

> the risk of escalation and retaliation in cyberspace. . . . Such activities could have a destabilizing effect on international peace and security. We stress that the risk of interstate conflict as a result of ICT incidents has emerged as a pressing issue for consideration. (G7 Declaration 2017, 1)

Over the past few years, several national and international actors—NATO (Freedberg 2014), the UN Institute for Disarmament Research (UN Institute for Disarmament Research 2014), the UK government (UK Government 2014), and the US State Department (International Security Advisory Board 2014)—have remarked on the urgent need to define strategies to ensure cyber stability in the wake of an escalating trend of cyber-attacks. Paradoxically, state actors often play a central role in the escalation of cyber-attacks. State-run and state-sponsored cyber-attacks have been launched for espionage and sabotage purposes since at least 2003. Titan Rain (2003), the Russian attack against Estonia (2006) and Georgia (2008), Red October (2007), Stuxnet and Operation Olympic Game against Iran (2006–2012) are well-known examples. Famous cases also involve the Russian cyber-attack against a Ukrainian power plant,[11] Chinese and Russian infiltrations of US federal offices,[12] the Shamoon/Greenbag

---

[11] https://www.wired.com/2016/03/inside-cunning-unprecedented-hack-ukraines-power-grid/.
[12] https://www.nytimes.com/2016/12/13/us/politics/russia-hack-election-dnc.html?_r=0.

cyber-attacks on government infrastructures in Saudi Arabia,[13] and the campaign of state-run/-sponsored cyber-attacks targeting European digital infrastructures and services following the Russian invasion of Ukraine.[14]

This trend will continue, posing an increasingly severe risk of escalation. Developing deterrence strategies that can address the nature of cyberwarfare is crucial but insufficient to mitigate these risks. Here, AI plays a central role. It is crucial that these strategies be coupled with regulations of state behaviour that are enforced by an authority with teeth. This is a complex effort, but also a necessary one, when considering the extent to which digital societies depend on their digital assets. This chapter, along with Chapter 4, had the goal of showing how conceptual and ethical analyses can support this effort to define effective deterrence strategies that, coupled with appropriate regulation, will improve the stability of cyberspace and of digital societies that rely on it to function.

---

[13] https://www.symantec.com/connect/blogs/greenbug-cyberespionage-group-targeting-middle-east-possible-links-shamoon.

[14] https://www.ncsc.gov.uk/news/russia-behind-cyber-attack-with-europe-wide-impact-hour-before-ukraine-invasion.

6

Adversarial and Kinetic Uses of AI: The Definition of Autonomous Weapon Systems

1. Introduction

I now turn to the use of AI for adversarial and kinetic purposes. Here the discussion centres on autonomous weapons systems (AWS) and the moral permissibility and regulation of their use. The debate on the ethical and legal implications of AWS dates back to the early 2000s, with some proponents (Arkin 2009) defending the use of these systems and others calling for an outright ban (Sharkey 2008, 2010; Sparrow 2007). The debate started in 2012, when the US Department of Defence (DoD) published an executive order on AWS (US Department of Defense 2012) and Human Rights Watch shared a report on the ethical and legal problems posed by AWS ("Losing Humanity" 2012). Since then, the debate has grown with contributions from scholars, military, and policy experts and with the involvement of the International Committee of the Red Cross (ICRC), the UN Institute for Disarmament Research (UNIDIR), and the UN Convention on Certain Conventional Weapons (CCW), which established a Governmental Group of Experts (GGE) to discuss emerging technologies in the area of lethal autonomous weapon systems (LAWS).

The debate remains deeply polarised as to whether the use of AWS is ethically acceptable and legally sound (I shall return to this point in Chapter 8). However, there is at least consensus on the

The Ethics of Artificial Intelligence in Defence. Mariarosaria Taddeo, Oxford University Press.
© Oxford University Press 2024. DOI: 10.1093/oso/9780197745441.003.0006

ethical and legal aspects that should be considered in making this call: respect for human dignity, for IHL, and for international stability. IHL is central to this debate, as there is consensus that AWS can be deployed only insofar as they abide by the IHL principles of necessity, proportionality, and distinction. These principles are uncontroversial; what is problematic is understanding whether, and to what extent, autonomous artificial agents enabling AWS can comply with them.[1] For example, respecting the principle of distinction for AWS is problematic given that—at least in its current state of development—autonomous artificial agents are unable to analyse the context in which they operate with the precision necessary to distinguish what or who is a legitimate target (Sharkey 2010, 2016; Amoroso and Tamburrini 2020).

The IHL principles define operational requirements that, if not met by current models of AWS, could at least in principle be met in the future by more refined AWS. More fundamental problems emerge when considering AWS and human dignity. In this case the questions is *how* a person is killed or injured. That is, the focus here is on the process through which the decisions to kill or injure a human are made: if the decision is taken by a machine, then the human dignity of those targeted is violated (Asaro 2012; Docherty 2014; A. Sharkey 2019; Johnson and Axinn 2013; Sparrow 2016; O'Connell 2014; Ekelhof 2019). The impact of the use of AWS on human dignity is independent of the level of sophistication of the technology, as the question here concerns the legitimacy of delegating the decision on the use of (lethal) force to machines (Lieblich and Benvenisti 2016). It questions whether delegating this decision is compatible with the values upheld by our societies and refers to the notions of humanity and public conscience, which are central to legitimacy of any weapons, not only AWS. As the ICRC report stresses:

---

[1] See Blanchard and Taddeo 2022b; on proportionality, Blanchard and Taddeo 2022c; and on necessity, Blanchard and Taddeo 2022a, in application to AWS.

ethical decisions by States, and by society at large, have preceded and motivated the development of new international legal constraints in warfare, including constraints on weapons that cause unacceptable harm. In international humanitarian law, notions of humanity and public conscience are drawn from the Martens Clause. (International Committee of the Red Cross 2018, 1)

Ultimately, problems relating to human dignity refer to human agency—the decisions and actions that we should or should not delegate to a machine, and the moral responsibilities linked to this agency and to the decision to use force. Ascribing moral responsibility for the actions performed by AI systems has proved to be problematic in many domains, and the case of AWS is not an exception. As I have argued elsewhere (Taddeo et al. 2021b), the responsibility gap is problematic in *all* the categories of use of AI within the defence domain, but this gap is particularly worrying when considering the adversarial and kinetic uses of AI given the high stakes involved (Sparrow 2007).

Questions also arise with respect to the impact of AWS on international stability. Some have argued that AWS may lower the barriers to warfare leading to an increased incidence of war and hamper international stability (Enemark 2011; Brunstetter and Braun 2013). For instance, the widespread use of AWS might allow decision-makers to wage wars without the need to overcome the objections of military personnel or of a democratic populace more broadly (Steinhoff 2013b; Heyns 2014). In the same vein, asymmetric warfare resulting from an aggressor using AWS may lead to the weaker side resorting to insurgency and terrorist tactics more often (Sharkey 2012a, 2012b). Because terrorism is generally considered to be a form of unjust warfare (or, worse, an act of indiscriminate murder), deploying AWS may lead to a greater incidence of unjust violence.

Scholarly and policy efforts focusing on these topics have grown over time. However, more than a decade after the DoD executive order and the Human Rights Watch report mentioned above, a shared international approach to address these problems has yet to be defined. The reasons for this failure range from lack of political will, competing interests at the international level, and defence postures, all of which is compounded by a lack of a shared understanding of AWS and of their key features and related ethical and legal implications. As stressed in a UNIDIR report, "proponents and opponents of AWS will seek to establish a definition that serves their aims and interests. The definitional discussion will not be a value-neutral discussion of facts, but ultimately one driven by political and strategic motivations" (UNIDIR 2017, 22). Indeed, the analysis proposed in this chapter identified 12 definitions of AWS proposed by state actors or key international actors like the ICRC and NATO. The definitions focus on different aspects of AWS and hence lead to different approaches to address the ethical and legal problems posed by these weapons systems. The lack of consistency in defining AWS is detrimental in terms of both fostering a common understanding of AWS and facilitating agreement on regulations of their use and, indeed, on whether AWS are to be used at all. This becomes evident when considering the work of the CCW/GGE. Table 6.1 summarises the key points of the discussion of this group between 2014 and 2019. It shows that while there is a consensus on the key aspects of AWS and on the ethical problems that they pose, a shared definition, and therefore a shared understanding of AWS and of what aspects present the most pressing ethical and legal problems, is still lacking. Consider for example, how the points reported in Table 6.1 often conflate AWS with LAWS and the related ethical and regulatory problems.

In this chapter, I offer a comparative analysis of existing definitions of AWS with the goal of identifying the different approaches that underpin them, their similarities and differences, as well as their limitations, in section 2. I then draw from this

Table 6.1 Key points of the discussions held at the CCW/GGE between 2014 and 2019

CCW/ GGE	2014	"Many interventions stressed the fact that, even if the elaboration of a definition was premature, some key elements appeared as pertinent to describe the concept of autonomy for LAWS, for example the capacity to select and engage a target without human intervention. Some experts highlighted the fact that autonomy should be measurable and should be based on objective criteria such as capacity of perception of the environment, and ability to perform pre-programmed tasks without further human action. Many interventions stressed that the notion of meaningful human control could be useful to address the question of autonomy. Other delegations also stated that this concept requires further study in the context of the CCW. The concept of human involvement in design, testing, reviews, training and use was discussed. The notion of predictability was also underlined by some delegations as a key issue" (Simon-Michel 2014, 4).
	2017	"The need to improve shared understanding of autonomous weapon systems was recognised. The elaboration of a working definition of LAWS, without prejudice to the definition of systems that may be subject to future regulation, was encouraged. Consideration was given to the scope of a possible definition, including questions of systems already deployed, defensive versus offensive weapons, and the distinction between fully and semi-autonomous systems. The view that it was premature or unhelpful to begin work on definitions was also put forward" (Korpela 2017, 7).
	2018	"Technical characteristics related to self-learning (without externally-fed training data) and self-evolution (without human design inputs) have to be further studied. Similarly, attempting to define a general threshold level of autonomy based on technical criteria alone could pose difficulty as autonomy is a spectrum, its understanding changes with shifts in the technology frontier, and different functions of a weapons system could have different degrees of autonomy.... In the context of the CCW, a focus on characteristics related to the human element in the use of force and its interface with machines is necessary in addressing accountability and responsibility" (Convention on Certain Conventional Weapons 2018, 5).

(continued)

Table 6.1 Continued

2019	"On the agenda item 5 (b) 'Characterization of the systems under consideration in order to promote a common understanding on concepts and characteristics relevant to the objectives and purposes of the Convention' the Group concluded as follows: (a) The role and impacts of autonomous functions in the identification, selection or engagement of a target are among the essential characteristics of weapons systems based on emerging technologies in the area of lethal autonomous weapons systems, which is of core interest to the Group; (b) Identifying and reaching a common understanding among High Contracting Parties on the concepts and characteristics of lethal autonomous weapons systems could aid further consideration of the aspects related to emerging technologies in the area of LAWS.... (b) Different potential characteristics of emerging technologies in the area of lethal autonomous weapons systems, including: self-adaption; predictability; explainability; reliability; ability to be subject to intervention; ability to redefine or modify objectives or goals or otherwise adapt to the environment; and ability to self-initiate" (UN GGE CCW 2019, 5).

analysis to identify essential aspects of AWS —autonomy, learning capabilities of AWS, human control, and purpose of use—and offer a definition that provides a value-neutral ground to facilitate efforts to address the relevant ethical and legal problems, in section 3. I conclude the chapter in section 4.

Before moving forward with the analysis, I should clarify that, for the purposes of this chapter, I focus on AWS and consider LAWS a subset of this category. That is, LAWS are AWS with a specific purpose of use, that is, deploying lethal force, in addition to the wider set of purposes of the use of AWS, for example, anti-materiel, damage, and destruction. This is important because the ethical problems related to AWS, including issues of control, responsibility, and predictability apply a fortiori when considering LAWS. At the same time, however, LAWS pose specific ethical problems related to the lethal purpose of their use, for example, respect for human dignity and for military virtue.

2. Definitions of Autonomous Weapon Systems

Table 6.2 lists 12 definitions of AWS or LAWS provided by States and international organisations. [2]

This plethora of definitions encroaches upon international debate on the ethical and legal implications of AWS. For example, it has been reported[3] that as of August 2020, 30 states had declared their endorsement of a pre-emptive ban on AWS. However, without a shared understanding of what AWS are, it is hard to identify what systems should be banned, let alone enforce such a ban. China offers a good example of the case in point. Roberts et al. (2020) highlight that Chinese military officials have expressed concerns about the use of AI for kinetic and aggressive purposes and that these concerns have motivated Chinese support for restricting the use of AWS, as expressed at the Fifth Convention on CCW, and, in the more recent call, for banning the use of LAWS. However, the definition of autonomy embraced by China is extremely narrow, as it focuses only on *fully* autonomous weapons and leaves unaddressed AWS that may have lower levels of autonomy (Kania 2018b).

A number of other definitions also focus on full autonomy. The UK definition centres on fully autonomous systems that are "capable of understanding higher-level intent and direction". The UK is out of step for its primary focus on the intention of the system,

---

[2] NATO offers a definition of *autonomous systems* and not specifically of AWS. Nonetheless, I include it here insofar as it refers to identifying characteristics of AWS.

[3] Human Rights Watch, "Stopping Killer Robots", August 10, 2020, https://www.hrw.org/report/2020/08/10/stopping-killer-robots-country-positions-banning-fully-autonomous-weapons-and.

Table 6.2 Twelve definitions of AWS and LAWS as provided by states or international organisations between 2012 and 2020

State Organisation	Date	Definition
Canada	2018	"Systems with the capability to independently compose and select among various courses of action to accomplish goals based on its [information] and understanding of the world, itself, and the situation." Note: while Canada has no official definition, this is the definition used by the Department of National Defence (Department of National Defence 2018; see also Shapiro 2019).
China	2018	"LAWS should include but not be limited to the following 5 basic characteristics. The first is lethality, which means sufficient pay load (charge) and for means to be lethal. The second is autonomy, which means absence of human intervention and control during the entire process of executing a task. Thirdly, impossibility for termination, meaning that once started there is no way to terminate the device. Fourthly, indiscriminate effect, meaning that the device will execute the task of killing and maiming regardless of conditions, scenarios and targets.
		Fifthly evolution, meaning that through interaction with the environment the device can learn autonomously, expand its functions and capabilities in a way exceeding human expectations" (China 2018, 1). Note: This definition differs from the definition set out by the People's Liberation Army in 2011: "[LAWS are] a weapon that utilizes AI to automatically pursue, distinguish, and destroy enemy targets; often composed of information collection and management systems, knowledge base systems, assistance to decision systems, mission implementation systems, etc." (Kania 2018b).
France	2016	"Lethal autonomous weapons are fully autonomous systems. LAWS are future systems: they do not currently exist.... LAWS should be understood as implying a total absence of human supervision, meaning there is absolutely no link (communication or control)

Table 6.2 Continued

State Organisation	Date	Definition
		with the military chain of command. . . . The delivery platform of a LAWS would be capable of moving, adapting to its land, marine or aerial environments and targeting and firing a lethal effector (bullet, missile, bomb, etc.) without any kind of human intervention or validation. . . . LAWS would most likely possess self-learning capabilities" (République Française 2016, 1–2). "Given the complexity and diversity of environments (particularly in urban areas) and the difficulty of building value-laden algorithms capable of complying with the principles of international humanitarian law (IHL), a LAWS would most likely possess self- learning capabilities, since it seems unrealistic to pre-program all the scenarios of a military operation. This means, for instance, that the delivery system would be capable of selecting a target independently from the criteria that have been predefined during the programming phase, in full compliance with IHL requirements. With our current understanding of future technological capacities, a LAWS would therefore be unpredictable" (République Française 2016, 2).
Germany	2020	"LAWS [are] weapons systems that completely exclude the human factor from decisions about their employment. Emerging technologies in the area of LAWS need to be conceptually distinguished from LAWS. Whereas emerging technologies such as digitalization, artificial intelligence and autonomy are integral elements of LAWS, they can be employed in full compliance with international law" (Federal Foreign Office 2020, 1).
International Committee of the Red Cross (ICRC)	2016	"Any weapon system with autonomy in its critical functions. That is, a weapon system that can select (i.e. search for or detect, identify, track, select) and attack (i.e. use force against, neutralize, damage or destroy) targets without human intervention" (International Committee of the Red Cross 2016, 1).

(continued)

Table 6.2 Continued

State Organisation	Date	Definition
Israel	2018	"In Israel's view, the shared starting point for this discussion must be that all weapons, including LAWS, are and will always be utilized by humans. We should stay away from imaginary visions where machines develop, create or activate themselves— these should be left for science-fiction movies. As far as terminology is concerned, that means that LAWS should not be regarded as "deciding" anything. Humans are always those who decide, and LAWS are decided upon" (Yaron 2018, 2).
NATO[1]		"Automated system: a system that, in response to inputs, follows a predetermined set of rules to provide a predictable outcome." "Autonomous system: a system that decides and acts to accomplish desired goals, within defined parameters, based on acquired knowledge and an evolving situational awareness, following an optimal but potentially unpredictable course of action" (NATO 2020, 16).
Norway	2017	"Norway has not yet concluded on a specific legal definition of the term 'fully autonomous weapons systems'. Generally speaking, however, in using the term, we refer to weapons that would search for, identify and attack targets, including human beings, using lethal force without any human operator intervening. These must be distinguished from weapons systems already in use that are highly automatic, but which operate within such tightly constrained spatial and temporal limits that they fall outside the category of 'fully autonomous weapons'" (Norway 2017, 1).
Switzerland		"Weapons systems that are capable of carrying out tasks governed by IHL in partial or full replacement of a human in the use of force, notably in the targeting cycle" (Switzerland 2016, 2).
The Netherlands	2017	"A weapon that, without human intervention, selects and engages targets matching certain predefined criteria, following a human decision to deploy the weapon on the understanding that an attack, once launched, cannot be stopped by human intervention" (The Netherlands 2017, 1).

Table 6.2 Continued

State Organisation	Date	Definition
United Kingdom[1]	2018	"An autonomous system is capable of understanding higher-level intent and direction. From this understanding and its perception of its environment, such a system is able to take appropriate action to bring about a desired state. It is capable of deciding a course of action, from a number of alternatives, without depending on human oversight and control, although these may still be present. Although the overall activity of an autonomous unmanned aircraft will be predictable, individual actions may not" (Ministry of Defence 2018a, 13).
	2016	"UK understands such a system [fully autonomous LAWS] to be one which is capable of understanding, interpreting and applying higher level intent and direction based on a precise understanding and appreciation of what a commander intends to do and perhaps more importantly why.... Critically, this understanding is focused on the overall effect the use of force is to have and the desired situation it aims to bring about.
		From this understanding, as well as a sophisticated perception of its environment and the context in which it is operating, such a system would decide to take—or abort—appropriate actions to bring about a desired end state, without human oversight, although a human may still be present. The output of such a system could, at times, be unpredictable—it would not merely follow a pattern of rules within defined parameters" (Foreign & Commonwealth Office 2016, 2).
US Department of Defense	2012	"A weapon system that, once activated, can select and engage targets without further intervention by a human operator. This includes human-supervised autonomous weapon systems that are designed to allow human operators to override operation of the weapon system, but can select and engage targets without further human input after activation" (US Department of Defense 2012, 13–14).

[1] The UK adopted the NATO definition of autonomous systems, but it did not abandon any of its previous definitions.

while its international partners focus on human (non)intervention with the system (Select Committee on Artificial Intelligence 2018, 105). This point has been affirmed in various meetings of the UN GGE and in a report by the UK House of Lords' Select Committee on Artificial Intelligence.[4] The definition refers to cognitive capabilities that AI systems do not possess currently and are unlikely to gain in the future (Floridi 2014; Wooldridge 2020). Indeed, "capable of understanding higher-level intent and direction" defines an atypically high threshold for what is to be considered autonomous. France's definition is provided in the same vein, and it explicitly mentions that AWS according to its definition "do not currently exist".

The French approach has the effect of informing future directions of technological innovation by indicating limits to possible uses of AI technologies. In doing so, it may enable regulators to gain an advantage over technological innovation. But this approach rests on a paternalistic view of the role of regulations and the regulator, which is problematic *per se* and may have the undesired effect of hampering technological innovation. When considering AWS specifically, defining the governance of these systems by focusing on futuristic scenarios is detrimental for two reasons. First, focusing on systems that are not currently developed or whose characteristics are technologically unfeasible diverts focus from pressing ethical and legal problems posed by existing AWS and those that may be used in the foreseeable future. Second, it undermines regulations and declarations about banning AWS, insofar as these

---

[4] Select Committee on Artificial Intelligence 2018, 105. On April 24, 2019, Lord Browne tabled a question in the House of Lords asking what representations the Government had received from the MoD regarding the recommended that the UK align its definition of AWS with that of international partners. The Government noted that it had received some representations but nevertheless pointed to the fact that "the UN Convention on Certain Conventional Weapons Group of Government Experts on Lethal Autonomous Weapons Systems is yet to achieve consensus on an internationally accepted definition or set of characteristics for autonomous weapons" (House of Lords 2019).

refer to hypothetical AWS with features that current and foreseeable systems do not have, for example understanding and intent. In this case, the implication is that official declarations about banning AWS refer to systems that do not exist yet and leave unaddressed other systems currently being used. For example, the NGO Article36 stressed that statements made by the UK such as "we have no plans to develop or acquire such weapons" – as per UK definition of AWS-- "could appear progressive without actually applying any constraint on the UK's ability to develop weapons systems with greater and greater autonomy" (Article36 2018, 1). Indeed, the high threshold established by the UK to identify AWS will, if unchanged, permit the UK the use of AWS insofar as these do not show "understanding higher-level intent and direction" (see Table 6.2). The problem in this case is conceptual: the restrictive definition of AWS does not enable the correct categorisation of these systems, which are autonomous, but which do not meet the high threshold posed by the UK definition. These systems either fall into a grey area between both categories or are mistakenly lumped into the more familiar category of automatic systems, missing opportunities to consider and address the ethical and legal problems that they pose.

To avoid these limitations, it is important to define AWS by focusing on their characterising aspects—for example, autonomy—and describe them according to a scientific and technological understanding. In this way, a definition can offer a rigorous tool to identify AWS and avoid focusing on unsubstantiated characteristics of these systems. The goal of such a definition, as the ICRC states, is that it

encompasses some existing weapon systems, [and so] enables real-world consideration of weapons technology to assess what may make certain existing weapon systems acceptable—legally and ethically—and which emerging technology developments may raise concerns under international humanitarian law (IHL) and under the principles of humanity and the dictates of the public conscience. (International Committee of the Red Cross 2016, 1)

This is, for example, the driving rationale of the ICRC definition (see Table 6.2) and the outcome of the US definition, which considers autonomy on a function-based spectrum vis-à-vis human engagement so it can also encompass *existing* weapons systems (International Committee of the Red Cross 2016, 1; US Department of Defense 2012, 13–14).

When working towards an inclusive definition, however, it is also important to maintain some level of specificity to avoid too generic an approach that could generate confusion. This is the risk of the NATO definition (see Table 6.2). It is true that the definition is not meant to focus specifically on AWS but on *autonomous systems* in general, but it is too generic even for this purpose. For example, it refers to the "desired goals" of a system, leaving unspecified whether these are the political, organisational, strategic, or tactical goals or the specific goals that the system itself may have or acquire. Similarly, it refers to "situational awareness", but it is unclear whether this is meant to be an understanding of the immediate context of deployment of the system or of the wider strategic scenario.

From the analysis of the definitions reported in Table 6.2, four characteristics can be extracted as recurring most often in the definition of AWS, namely autonomy, learning capabilities, human intervention and control, and purpose of use. While these characteristics point in the right direction when considering what AWS are, for example they resonate with the definition of AI adopted by Taddeo, McCutcheon, and Floridi (2019) and Taddeo et al. (2021b) of a form of autonomous, self-learning agency, the way in which they are described is, at times, conceptually misleading. The next three sections analyse each of these characteristics in turn, clarifying their implications with respect to the ethical debate on AWS.

2.1. Autonomy, Intervention, and Control

Autonomy is a central element of all the definitions of AWS listed in Table 6.2. In some cases, it is assumed to mean the ability of a system to operate successfully without human intervention. The German definition, for example, mentions machines that "completely exclude" humans from the decision-making process. In other cases, autonomy is conflated with a lack of human control. This is the case of the French definition, for instance, which defines LAWS as having "absolutely no link (communication or control) with the military chain of command" (République Française 2016, 1).

As we will see in section 3, conflating autonomy with the lack of human control is misleading both conceptually and operationally, given that a fully autonomous AI system can still be deployed under some form of meaningful human control.

The distinction between autonomy and control is important, as it brings three advantages. First, for conceptual clarity, we should not regard automation and human control as mutually exclusive concepts. Automation makes human direct intervention for the achievement of a given goal unnecessary, but this does not exclude forms human control. For example, monitoring the autonomous performances of a system to unplug it if it does something wrong. This is why the DoD Directive 3000.09 is correct in referring explicitly to "human-supervised autonomous weapons systems" (US Department of Defense 2012, 14) and in distinguishing these from "semi-autonomous weapon systems", whose autonomy is circumscribed to "engagement related functions" that depend on a human operator for the target selection.

Distinguishing autonomy from control brings a second advantage as it future-proofs the debate on AWS. Many of the problems posed by AWS do not concern the desirable level of autonomy of these systems, but the desirable level of human control over these systems. The decision about control is in many ways normative, insofar as it is not only defined by the technological affordances (e.g., how fast a system is) but also, and more importantly, by the decisions and tasks that should be delegated to machines without requiring human control. Separating the two concepts enables a focus on normatively desirable forms of control, irrespective of the level of autonomy that these machines may acquire someday.

The third advantage of distinguishing autonomy from control is that it pre-empts approaches that use the lack of existing examples of fully autonomous AWS to conclude that there are no real questions with respect to the control of AWS as humans are still involved in or on the loop. This an approach oftern underpinning attempts to avoid discussing the need to regulate/ban the use of AWS (Russian Federation 2017, 2).

2.2. Learning Capabilities

Of the 12 definitions considered in this chapter, only the French and the Chinese definitions stress the learning capabilities of AWS as a key characteristic. The lack of focus on learning capabilities in the definition of AWS is problematic, given this is a key feature of AI technologies. Of course, AWS can function without learning capabilities. For example, they may rely on rule-based programming that enables an autonomous reaction to environmental triggers but does not allow for planning new behaviours when the environment changes. One can imagine a sensor detecting an incoming object and the algorithm triggering a system response of the system, for example, fire to destroy the object.

As the technology develops, systems based on rule-based algorithms are being replaced by AI-based systems. Military institutions are investing in AI for a wide range of applications. For example, significant efforts are already underway to harness developments in image, facial, and behaviour recognition using AI and ML techniques for intelligence gathering and "automatic target recognition" to identify people, objects, or patterns (the reader may recall the discussion on AI for intelligence analysis of Chapter 3).

Disregarding learning capabilities in the definitions of AWS leads to disregarding a key characteristic of these systems and hinders debate on their ethical and legal implications. Crucially, as we saw in Chapter 1, AI capabilities pose questions with respect to their predictability, and hence reliability; with respect to the attribution of responsibilities of the actions that these systems perform; as well as with the implementation of meaningful forms of control. The French definition stresses that learning capabilities would be necessary to adapt to the complexity of operation scenarios that cannot be foreseen and thus "pre-programmed" in the system. It also stresses that this means

> that the delivery system would be capable of selecting a target independently from the criteria that have been predefined during the programming phase, in full compliance with IHL requirements. With our current understanding of future technological capacities, a LAWS would therefore be *unpredictable*. (République Française 2016, 2, emphasis added)

The ICRC highlights a similar point: "The application of AI and machine learning to targeting functions raises fundamental questions of inherent unpredictability" (International Committee of the Red Cross 2018, 2). Learning capabilities—leading to unpredictability of outcomes—pose problems with respect to Article 36 of Additional Protocol I to the Geneva Conventions on weapons review:

> [f]rom a technical perspective, any system that continues to learn while deployed is constantly changing. It is not the same system it was when deployed or verified for deployment. Some have raised questions about the legality of adaptive systems, particularly in regards to States' Article 36 obligations. (UNIDIR 2017, 10)

This is crucial, as noted by the ICRC:

> [t]he ability to carry out [an Article 36] review entails fully un-
> derstanding the weapon's capabilities and foreseeing its effects,
> notably through testing. Yet foreseeing such effects may become
> increasingly difficult if autonomous weapon systems were to be-
> come more complex or to be given more freedom of action in
> their operations, and therefore become less predictable. (ICRC,
> as reported in UNIDIR 2017, 26)

For both ethical and legal reasons, therefore, a focus on the learning
capabilities of AWS is essential. It is the nature of the learning
process that raises both significant opportunities and challenges
and sets AI-enabled systems apart from highly automated rules-
based systems. Learning capabilities qualify the latest and future
generations of AWS. Focusing on them allows for further clarification of the distinction between automatic and autonomous systems (more on this in section 3); and for identifying the source of
a number of key ethical and legal implications of AWS. This is why
it is important that definitions of AWS mention these capabilities
explicitly. It is problematic that even the two most comprehensive
definitions of AWS—from the US and the ICRC—fail to grasp this
point, missing the opportunity to cast light on a key element of
these systems.

2.3. Purpose of Deployment

Most of the definitions in Table 6.2 define the purpose of deployment of AWS only implicitly, by making reference to "weapons" and
stating that AWS are deployed in kinetic contexts. This indicates
some form of destructive (whether anti-materiel or lethal) use of
these systems. However, it is important to understand the range of

possible uses with greater precision, for example considering the specific tasks that AWS may undertake within the context of kinetic operations.

Of the definitions in Table 6.2, four (Canada, Israel, Germany, and UK) do not mention explicitly any specific purpose of deployment. The kinetic outcome of the use of AWS is assumed in this case, leaving undefined, for example, whether AWS will be used for deliberate or dynamic targeting. Of the other eight definitions, one (NATO) does not mention any specific purpose (it should be stressed, however, that the NATO definition is of autonomous systems in general and not of AWS), and the remaining definitions refer to AWS as deploying lethal force (China and France) or more specifically to select and engage targets (whether humans or nonhumans) to be neutralised, damaged, or destroyed (ICRC, Norway, Switzerland, Netherlands, US).

All the definitions leave unaddressed the specific steps of the tasks that are delegated to machines. These steps, however, are key when considering AWS and the ethical and legal implications of their use. Consider, for example, Roff's (2014) criticism of the US definition, that the meaning of "select" in "select and engage" is unclear, insofar it is not clear whether this also includes the detection of targets (Conn 2016). As she argues, if detection is not included in the definition, then we may assume that this is carried out by a human, thereby obviating important ethical (and technical) questions.

Roff's criticism highlights the complexity of these tasks and of the processes underpinning the decision to deploy force. Consider, for example, the steps underpinning targeting decisions as described by Ekelhof and Persi Paoli (2021). They outline a complex process that extends across the decision and command chain when considering AWS. The process includes tasks and decisions spanning the tactical, operational, strategic, and political levels, which are often interlinked.

The complexity of the process requires a more specific approach when considering the tasks performed by AWS. This is achieved in three ways, by specifying explicitly

- the purposes of deployment of these systems and the destructive goals (whether lethal or not),
- the steps in the process of exerting force which sit within the remit of the AWS,
- the level of human control under which it operates. The outcome of the ethical and legal analyses of AWS depends on these specifications.

3. A Definition of AWS

At this point, I will offer a value-neutral definition of AWS. In doing so I have a double goal of defining the key characteristics that permit the identification of AWS; and describing these characteristics and how they are related, like automation and control, and how they differ, for example automatic versus autonomous. To do so, I consider autonomy, learning capabilities, and control as characteristics that can each be mapped on a continuum. AWS can have each of these characteristics to a greater or lower degree. I am also inclusive with respect to the set of possible purposes of deployment, with the aim of clarifying what the range may be. Identifying the combination of the different levels of these characteristics and purposes of use (if any at all) that may meet particular ethical and legal requirements is the task of ethical analyses (I leave it to the following chapters). With this approach in mind, I define AWS as follows:

an AWS is an artificial agent that, at the very minimum, is able to change its own internal states to achieve a given goal or goals within its dynamic operating environment and without the direct intervention of another agent (i.e., it is an automated artificial agent), and may also be able to change its own transition rules to adapt to the environment or refine its behaviour (i.e., it has learning capabilities) without the direct intervention of another agent, and which is deployed to exert kinetic force against a physical object or human and, to this end, is able to identify, select, and attack the target without the direct intervention of another agent. Once deployed, AWS can operate with or without human control.

The next subsections will unpack this definition by focusing on the concepts of autonomy, learning capabilities, and control. The purposes of deployment are less conceptually problematic, and thus I will not delve into them. It is important, however, to say here that the purposes of deployment have been identified as being those directly related to the goal, that is, exerting force. Selecting targets and engaging them (whether deliberate or dynamic) are directly linked to the purpose of deploying force. Hence, a system whose selecting and attacking functions are autonomous, but which is directed by another agent(s) for all its other purposes of use, for example, mobility, would still be considered an AWS.

3.1. Autonomous, Self-Learning Weapons Systems

A key question underpinning the definition of AWS is the distinction among *automatic*, *automated*, and i systems. Especially the distinction between automated and autonomous can prove to be difficult when considered from an ethical LoA. An ICRC report, for example, stresses, "[t]here is no clear technical distinction between automated and autonomous systems, nor is there universal agreement on the meaning of these terms" (International Committee of the Red Cross 2019, 7). In a similar vein, Joint Concept Note 1/

18, "Human-Machine Teaming", published by the UK Ministry of Defence in 2018, starts by remarking, "There is no clear, definable and universally agreed boundary between what constitutes automation and what is autonomous . . . because the assessment of autonomy and the term's use is subjective and contextual" (Ministry of Defence 2018b, 57). One may agree that the distinction between automation and autonomy is blurred, but this is not because the assessment of autonomy of artificial agents is subjective or context-dependent. Within the field of computer science, and particularly of *agent theory* there is a clear understanding of the differences between these concepts (Wooldridge and Jennings 1995; Castelfranchi and Falcone 2003).

Let us consider automatic agents first. These are agents whose actions are predetermined and will not change unless acted upon by pre-selected triggers and/or human intervention. Automatic agents are not teleological; they do not pursue a goal, but simply react to an external trigger. In this sense, they are "causal entities" (Castelfranchi and Falcone 2003). A landmine falls squarely in this category, for its action is causally determined by a specific trigger, such as someone stepping on it. AWS do not belong to this category insofar as their behaviour is predetermined.

AWS execute tasks to achieve goals (teleological agents); they can adjust their actions on the basis of the feedback that they receive from the environment (automated agents), may be able define plans (heuristic agents) to achieve their goals, and may also be able to refine their behaviour in response to changes in the environment (learning agents). At this point, we can consider AWS as systems that at the very least are automated, teleological artificial agents, but we can be more specific and go a step further.

For the purposes of the definition, it is important to consider what the minimum requirements are for an artificial agent to be autonomous. To do so we will refer to the definitions of autonomous artificial agents provided by Castelfranchi and Falcone (2003) and Floridi and Sanders (2004).

According to Castelfranchi and Falcone, autonomous agents enjoy the following properties:

> [t]heir behaviour is *teleonomic*: it tends to certain specific results due to internal constraints or representations, produced by design, evolution, or learning . . . ; they do not simply receive an input—not simply a force (energy) but information—but *they (actively) "perceive" and interpret their environment* and the effects of their actions; . . . *they orient themself towards the input*; in other words, they define and select the environmental stimuli; . . . *they have "internal states"* with their own exogenous and endogenous evolution principles, and their behaviour also depends on such internal states. (2003, 105)

The internal state of an artificial agent can be described as the configuration of the agent (for example, the values and the weights of a neural network at a specific moment in time) when it is performing a given operation. Internal states are key in the definition of autonomy, insofar as the transition between states corresponds to a change of behaviour of the system. How the transition is determined defines the difference between automated and autonomous systems. Indeed, internal states are also key to the definition offered by Floridi and Sanders, in which autonomous artificial agents enjoy three characteristics:

> "*Interactivity* means that the agent and its environment (can) act upon each other. Typical examples include input or output of a value, or simultaneous engagement of an action by both agent and patient—for example gravitational force between bodies.
>
> *Autonomy* means that the agent is able to change state without direct response to interaction: it can perform internal transitions to change its state. . . .

Adaptability means that the agent's interactions (can) change
the transition rules by which it changes state. This property
ensures that an agent might be viewed, at the given LoA, as
learning its own mode of operation in a way which depends
critically on its experience" (2004, 357).

The ability of an artificial agent to change its internal states without
the direct intervention of another agent marks the line between au-
tomatic/automated and autonomous. A rule-based artificial system
and a learning one both qualify as autonomous following this
criterion.

As mentioned in section 2.1, learning capabilities are an increasingly common characteristic of AWS. It is both the characteristic
that underpins their potential for dealing with complex, fast-paced
scenarios and the one that leads to unpredictability, lack of transparency, lack of control, and responsibility gaps relating to the use
of these agents. Thus, it is important to include these capabilities
in the definition of AWS, to offer a clear and technical specification
of these capabilities to avoid anthropomorphising these agents,
and to set a clear threshold below which one can say that an agent
has no learning capabilities. This is why in the definition of AWS
proposed above I refer to an artificial agent endowed with some
abilities for changing *its transition rules* to perform successfully in a
changing environment.

3.2. Human Control

The definition of AWS provided in section 3 refers to human control
as a mode of deploying AWS and not as a defining characteristic.
This is because the autonomy of AWS is not defined with respect
to human control but with respect to the *intervention* of another
agent on the AWS. There are different forms of control (Tsamados

and Taddeo 2023). For example Amoroso and Tamburrini identify three:

> [f]irst, the obligation to comply with IHL entails that human control must play the role of a fail-safe actor, contributing to prevent a malfunctioning of the weapon from resulting in a direct attack against the civilian population or in excessive collateral damages. Second, in order to avoid accountability gaps, human control is required to function as accountability attractor, i.e., to secure the *legal* conditions for responsibility ascription in case a weapon follows a course of action that is in breach of international law. Third and finally, from the principle of human dignity respect, it follows that human control should operate as a moral agency enactor, by ensuring that decisions affecting the life, physical integrity, and property of people (including combatants) involved in armed conflicts are not taken by non-moral artificial agents. (2020, 189)

One may disagree with this taxonomy or consider control better defined at a different LoA, for example focusing only on the technical specifications of AWS. However, the relevant literature converges on considering control of AWS as dynamic, multidimensional, and situation dependent, and as something that can be exercised by focusing on different aspects of the human-machine team. For example, the Stockholm International Peace Research Institute and the ICRC identify three main aspects of human control of weapon systems: the weapon system's parameters of use, the environment, and human-machine interaction (Boulanin et al. 2020). More aspects can also be considered. Boardman and Butcher (2019) suggest that control should not just be meaningful but also appropriate, insofar as it should be exercised in such a way as to ensure that the human involvement in the decision-making process remains significant without impairing system performance.

The discussion about what constitutes meaningful human control of AWS and whether it can be exerted in an appropriate way does not fall within the scope of this chapter, as our goal here is to identify the key characteristics of AWS more than the normative conditions for their design, development, and deployment. However, to the extent that this analysis sheds light on these characteristics and their relation, it is important to stress that human control is not antithetical to the autonomy of AWS and can be exerted over AWS at different levels, from the political and strategic decisions to deploy AWS to the kinds of task delegated to them. The question is which form of control is ethically desirable and, assuming it is feasible, should therefore be considered by policymakers in designing a framework for the governance of AWS.

4. Conclusion

The debate on AWS is shaped by strategic, political, and ethical considerations. Competing interests and values polarise the debate, while politically loaded definitions of AWS undermine efforts to identify legitimate uses and to define relevant regulations. These efforts are further hindered when conceptual confusion is added to the picture.

In this chapter, my goal was to overcome the conceptual confusion around AWS, through a comparative analysis of current definitions, and to offer a new and value-neutral definition of these systems. The comparative analysis of the current definitions of AWS helps in identifying key points, for example, the distinction between automatic and autonomous and between autonomy and control. It also highlights a serious gap in these definitions concerning the learning capabilities of these systems.

The definition I have offered is not informed by policy or strategic aims, nor does it include normative aspects. It has been designed with the sole purpose of enabling the identification of AWS and distinguishing these systems from other weapon systems, such as automatic systems. I believe that having a value-neutral definition of AWS will help academic and policy debates on this topic, as it offers a shared ground on which different views can be compared.

7

Taking a Moral Gambit: Accepting Moral Responsibility for the Actions of Autonomous Weapons Systems

1. Introduction

A key question when considering the use of AI in defence is the attribution of moral responsibility for the actions of AI systems. This question pertains to all the three categories of use of AI in defence, but it is extremely relevant when considering adversarial and kinetic uses, particularly of AWS. Only humans are morally responsible for the actions of AWS. This is because intentions, plans, rights, duties, praise, or punishment can only be attributed in a meaningful way to humans. Placing this moral responsibility on AWS, or in general on AI systems, would entail misplacing "causal accountability and legal liability regarding their mistakes and misuses. Robots could be blamed and punished instead of humans. And irresponsible people would dismiss the need for care in the engineering, marketing and use of robots" (Floridi and Taddeo 2018, 309). Consensus on placing moral responsibilities for the actions of AWS on human agents has grown to the extent that this position is now uncontroversial. For example, the UN GGE CCW identifies human responsibility as a key principle for the (possible) use of LAWS, stating "[h]uman responsibility for decisions on the use of weapons systems must be retained since accountability cannot be transferred to machines. This should be considered across the entire lifecycle of the weapon system" (UN GGE CCW 2019).

The Ethics of Artificial Intelligence in Defence. Mariarosaria Taddeo, Oxford University Press.
© Oxford University Press 2024. DOI: 10.1093/oso/9780197745441.003.0007

As the reader may recall from Chapter 1, human responsibility is mentioned explicitly in the ethical principles for the use of AI in the Recommendations on the Ethical Use of Artificial Intelligence by the US DoD, whose first principle stresses that "[h]uman beings should exercise appropriate levels of judgment and remain responsible for development, deployment, use, and outcomes of DoD AI systems" (DIB 2020a, 8). Similarly, in a report to the European Parliament, the Committee for Legal Affairs stated that "autonomous decision-making should not absolve humans from responsibility, and . . . people must always have ultimate responsibility for decision-making processes so that the human responsible for the decision can be identified" (Lebreton 2021). The consensus on this point is important, for it avoids the risk of anthropomorphising AWS and AI systems more broadly and saves time spent engaging with the literature discussing whether AWS can be morally responsible for their own actions. Insofar as these systems have no intentions and no understanding of the blame or praise that may follow their actions, AWS do not bear moral responsibility for their own actions.

The focus on human agents points the debate in the right direction. Nonetheless, it remains problematic to ascribe moral responsibilities to humans for the actions of AWS in a *meaningful* way (more on this presently). Yet, ascribing this moral responsibility to individuals—as opposed to institutions or legal entities—is a necessary condition for the deployment of AWS. As was stressed in the Nuremberg trials following World War II, "Crimes against international law are committed by men, not by abstract entities, and only by punishing individuals who commit such crimes can the provisions of international law be enforced" (International Military Tribunal [Nuremberg] 1947, 221). In the age of autonomous warfare, AWS may perform immoral actions, but it is only by holding morally responsible the individuals who design, develop, and deploy AWS that the morality of warfare can be upheld.

In this chapter, I will analyse the main contributions to the debate on the moral responsibility for AI systems in general, in sections 2 and 3. I then will delve into specific approaches to ascribing moral responsibility for AWS and consider their relative strengths and limitations in section 4. I will then offer my own contribution to the debate by focusing on the concepts of *meaningful moral responsibility* and *moral gambit*, in section 5, and then provide eight recommendations addressing how defence institutions can overcome the responsibility gap for the use of *non-lethal AWS*, in section 6. I conclude the chapter in section 7.

Before beginning the analysis three clarifications are necessary. First, in this chapter, I will focus on AWS according to the definition I gave in Chapter 6:

> [a]n AWS is an artificial agent that, at the very minimum, is able to change its own internal states to achieve a given goal or goals within its dynamic operating environment and without the direct intervention of another agent (i.e., it is an automated artificial agent), and may also be able to change its own transition rules to adapt to the environment or refine its behaviour (i.e., it has learning capabilities) without the direct intervention of another agent, and which is deployed to exert kinetic force against a physical object or human and, to this end, is able to identify, select, and attack the target without the direct intervention of another agent. Once deployed, AWS can operate with or without human control.

This is a value-neutral definition, so it has no other bearing on our analysis than to identify the set of AWS to which it applies. It is worth noting again that this definition of AWS differs from many others, insofar as it explicitly considers learning capabilities as a key feature of AWS. It is this characteristic that drives much of the discussion in this chapter.

This takes us to the second clarification. Following the above definition of AWS, lethal and non-lethal AWS are distinguished on the basis of the *purpose* of their use and not the effect of their use. A lethal AWS is used with the purpose of exerting lethal force—that is, resulting in death or permanent injury of human beings. A non-lethal AWS is used with the purpose of incapacitating human beings "without causing death or permanent injury" (Davison 2009, 1). As the reader may recall from Chapter 1, the outcome of AWS is predictable only to an extent. Thus it is conceivable that an AWS used for non-lethal purposes may produce lethal effects (Coleman 2015; Enemark 2008; Kaurin 2010, 2015; Heyns 2016a, 2016b). This is because there is "a potential disconnect between the intention behind the use of a weapon and the consequences thereof" (Enemark 2008b, 201). This scenario does not invalidate the distinction that I propose here. Rather, it exemplifies precisely the one of the questions that I address in this chapter; that is the attribution of moral responsibility for the actions of unpredictable systems. Part of the analysis of the moral responsibility of AWS that I propose here applies squarely to both LAWS and non-lethal AWS. I will distinguish between non-lethal AWS and LAWS whenever the analysis leads to a different outcome for each set of AWS.

The third clarification addresses the nature of the responsibility on which we will focus. Our goal is to understand how a human agent can take *meaningful moral* responsibility for the actions of AWS. This implies that moral responsibility is not ascribed nominally to human agents, for example because of their role or ranking, but bears in a justified and fair way on those who have played a key role in the realisation of the effects of using AWS. This also implies that a human agent accepts this moral responsibility, and the praise or blame that comes with it, as an individual (in a personal sense) and not as a member or representative of a defence organisation or of a professional body. The attribution of meaningful moral responsibility may underpin legal processes to attribute legal responsibility and define accountability and liability. However, while related to legal responsibility, moral responsibility differs from it. For example, as we shall see in the next section, the attribution of moral responsibility and the subsequent blame/praise requires causal and intentional connection between an action and effect. This is not necessarily the case when considering legal responsibility. Consider, for example, the concept of *faultless responsibility*, according to which punishment can be meted out even if no intention to commit wrong can be determined. This chapter focuses only on moral responsibility, and while it may provide the conceptual ground for attributing legal responsibility, it is not concerned with the latter.

With the conceptual space of our analysis outlined, we can now delve into the literature focusing on attributing moral responsibility for AI systems.

2. Moral Responsibility for AI Systems

The debate on the moral responsibilities for the actions of AWS hinges on the wider discussion of the moral responsibilities for the actions of AI systems. As mentioned above, there is growing consensus that these responsibilities rest with human agents, it is ascribing them correctly that remains problematic. This is because, under classic ethical approaches, for the attribution of moral responsibility to be justified and fair, agents must have a specific relation to their own actions and their consequences. This relation must satisfy all four of the following conditions:

- Intentionality condition: the agent has to have an intention to achieve a given effect (Kant and Borken 2019; Branscombe et al. 1996; Khoury 2018).
- The causality condition: there has to be a causal connection between the decision/action of the agent and its effects (Fischer and Ravizza 2000; Sartorio 2007; Shoemaker 2017).

- Consequence condition: the agent has to have an understanding of the effects of the decision/action, as well as of their moral value and the consequent blame/praise (Bentham 1789; Wallace 1998; Levy 2008; E. I. Kelly 2012).
- Choice condition: the agent has some degree of freedom that allows him/her/them to choose between different patterns of actions (Strawson 1962; 1962; Watson 1975; Nelkin 2011).[1]

Al four conditions are problematic to meet in AI systems, and problems with meeting these conditions underpin the *responsibility gap* (Matthias 2004; Floridi 2012). For example, the distributed nature of the design/development/deployment cycle and the lack of transparency of AI systems may make it hard or impossible to understand what actions caused a particular outcome. At the same time, a human agent involved in the AI life cycle may lack the necessary understanding of the technology, or of its purposes of use, or of its effects once deployed to be able to consider and assess the relevant consequences as defined by the consequent condition. The possibility of meeting the choice condition depends on the definition of freedom and choice. However, problems in meeting the last three conditions are contingent; there is nothing inherent to the use of AI that makes meeting these conditions impossible to be met. This is not the case with the intentionality condition, whose satisfaction is impossible because of the inherent limited predictability of AI systems. This is why I shall focus on this condition in the rest of this chapter.

Moral responsibility is often ascribed with the aim of distributing praise or blame to individuals for their morally good or evil actions.

---

[1] In the rest of this chapter, I will not focus on the choice condition, as in the relevant literature this refers to metaphysical understanding of determinism and freedom. The condition focuses on whether humans are or not fully determined and, thus, can choose among alternative pattern of actions. The answer as to whether an AI system meet this condition independent from the features and level of refinment of these systems and more related to the metaphysical view one has.

For this, intentionality of the agent is a key element to ensure that the allocation of moral responsibility is justified.

> It would be counterproductive to attribute responsibility, and hence allocate blame or praise, punishments or rewards, if the agents' actions were not intentional, because such attribution would then be *arbitrary* and *indistinguishable* from a mere random allocation, which would defy the purpose of blame or praise, punishments or rewards. (Floridi 2016, 4, emphasis added)

Under a classic ethical approach, lack of intentionality undermines the allocation of moral responsibility and of the praise/blame linked to it, even when the causal chain of events leading to a given outcome is clear. The behaviour of AI systems may not result directly from the intentions of the individual designers, developers, or deployers. This can be the case for two reasons: distributed actions and lack of predictability. AI systems can perform morally loaded actions that stem from a number of morally neutral actions—that is, individual actions performed by humans or other artificial agents that do not in themselves (alone) lead to specifically good or evil outcomes (Floridi 2016). We can imagine a network of agents involved in the design, development, and use of an AI system, each making morally neutral decisions, that, once coordinated at the network level, lead to a morally evil outcome.

This is what Floridi (2012) calls distributed morality. While the entire network can be held morally responsible for these actions, attributing blame/praise to any individual or group of agents in the network would be unjustified insofar as individual actions *per se* do not lead to any morally loaded outcome—even if they do once coordinated at the network level. Distributed morality is not unique to AI systems, but it is particularly relevant in the AI domain, where networks of involved agents and fragmentation of tasks among them are particularly widespread, making it difficult to identify intentionality for actions and attribute moral responsibility accordingly. At the same time, once deployed, some AI systems may develop new behaviours that were never intended by the human agents designing, developing, and deploying these systems, and for which they cannot be held morally responsible—given the lack of intentionality. To put it more precisely: "Intentionality is not closed under causal implication. . . . In the *direct* case of non-closure, it is not the case that, if Alice means to cause *a*, and *a* causes *b*, it follows that Alice means to cause *b*" (Floridi 2016, 4). This is relevant when considering AI systems that may develop unforeseen and unintended behaviour (i.e., the predictability problem). The predictability problem makes it problematic to meet the consequence condition. Insofar as the outcomes of the system that one deploys cannot be foreseen, it is not feasible for the agent to consider all possible consequences of this action and their moral value.

Lack of predictability is at the nexus of the breach of the intentionality condition. This is, thus, a key element to address when considering moral responsibility for AI. As the reader may recall from Chapter 1, predictability is impacted by a large set of variables: the technical features of the system, the characteristics of the context of deployment, the level at which the operator understands the way in which the system works, and, in the defence domain, the behaviour of the opponents. These variables may change and interact together in different ways, making it difficult to predict all possible actions that an AI system may perform and their effects. The limited predictability of a system's outcomes makes it difficult, if not impossible, to link the outcomes to the intentions of human agents designing, developing, and deploying these systems. In turn, this makes it impossible to ascribe moral responsibility for the actions of AI systems to human agents following the classic approach.

3. Collective and Faultless Distributed Moral Responsibility

Alternatives to the classic ethical approach to assigning human moral responsibility for the actions of AI systems have been proposed in the literature on collective (Corlett 2001; List and Pettit 2011) or distributed moral responsibility (Floridi 2012; 2016).

Approaches focusing on collective responsibility (A. Krishnan 2009; List and Pettit 2011) address the allocation of moral responsibility for actions taken by a group collectively, rather than by a member individually. Consider, for example, the analysis by List and Pettit:

> [a] group agent [i.e. a group of people, treated as a single, unitary agent] is fit to be held responsible for doing something to the extent it satisfies these requirements:
>
> First requirement. The group agent faces a normatively significant choice, involving the possibility of doing something good or bad, right or wrong.
>
> Second requirement. The group agent has the understanding and access to evidence required for making normative judgments about the options.
>
> Third requirement. The group agent has the control required for choosing between the options. (2011, 158)

This analysis hinges on the premise that a group expresses its will and acts consequently. As the authors continue:

> to satisfy the second condition . . . , a group agent must be able to form judgments on propositions bearing on the relative value of the options it faces—otherwise it will lack normative

understanding—and it must be able to access the evidence on related matters. (158)

According to this approach, the group is seen as a single homoge-
neous entity—one may think, for example, of a group of workers
striking or a group protesting in the street—that acts in a coordinated manner. This is problematic when considering the life cycle
of AI systems, insofar as this is quite distributed and involves heterogeneous agents, who may not know of the overall action that
the group is performing, let alone make normative judgments. By
assuming an intentionality at group level, this approach disregards
the impact that unintentional individual actions may have on the
actions performed by the group. Thus, to rely on this approach
would lead to unjust distribution of moral responsibility. As Corlett
puts it:

> a collective *behavior* is a doing or behavior that is the result of a
> collective, though not the result of its intentions. A collective *ac-*
> *tion* is caused by the beliefs and desires (wants) of the collective
> itself, whether or not such beliefs and desires can be accounted
> for or explained in individualistic terms. . . . I am concerned
> with whether or not it is justified to ascribe intentional action to
> conglomerates of a numerically larger sort such as (large) nations
> and (large) corporations. If such conglomerates are not inten-
> tional agents, then they are not proper subjects of moral respon-
> sibility attributions. (2001, 575–76, emphasis added)

I agree with Corlett: ascribing intentionality to the group without
being able to ascribe intentionality to all its agents undermines the
idea of holding a group morally responsible, for those who did not
share the intentions of the group would bear this responsibility
without a justification.

The distributed morality approach—by contrast with the collective approach—focuses on the attribution of responsibility for morally good or evil actions that arise from the convergence of different, independent, morally neutral, intention-less factors. This has been defined as *distributed faultless moral responsibility* (Floridi 2016). It refers to contexts in which, while it is possible to identify the causal chain of agents and actions that led to a morally good or bad outcome, it is not possible to attribute intent to achieve such an outcome to any of those agents individually, and, therefore, all the agents are held morally responsible for that outcome insofar as they are part of the network that determined it. According to this approach, to attribute moral responsibility, what one needs to show is that

> some evil has occurred in the system, and that the actions in question caused such evil, but it is not necessary to show exactly whether the agents/sources of such actions were careless, or whether they did not intend to cause them. (Floridi 2016, 8)

All the agents of the network are then held maximally responsible for the outcome of the network.

It is important to note that this approach allows for distributing moral responsibility among the human agents of a network but does not distribute punishment or reward for the actions of a system. Its goal is to establish a feedback mechanism that incentivises all the agents in the network to improve its outcomes—if all the agents are morally responsible, they may become more cautious and careful, and this may reduce the risk of unwanted outcomes. This element is of particular relevance when considering AWS. However, the distributed faultless moral responsibility model does not allow us to close the responsibility gap in a satisfactory way when considering AWS; because attributing blame or praise in the case of AWS is crucial.

4. Moral Responsibility for AWS: The Collective Moral Responsibility Approach

Attempts to overcome the responsibility gap specifically for AWS have already been proposed in the literature. Some approaches build on collective responsibility; others rely on the ideas of allocating responsibility along the chain of command or on distributed moral responsibility. In this section, I shall address these three approaches in turn, starting with those building on collective responsibility. Following the collective idea of moral responsibility, one could hold responsible the entire network of agents involved in the AI life cycle. For example, Taylor states: "I maintain, the organisation as a whole might be viewed as having control over the outcome and thus be properly held morally responsible" (2020, 327). This approach rests on the idea that collective responsibility is ascribed to groups of agents who share the intent to perform a given action (List and Pettit 2011). Indeed, it can be argued that those who work to design, develop, and deploy AWS share the intention to develop a system that can deploy (lethal) force in a specific way and within certain constraints, and hence bear moral responsibility for the actions that the system is intended to perform. For the private sector, for example, this responsibility is akin to the liability that technology providers have with respect to the possible failure of their products. For example, Schulzke stresses that

> to the extent that AWS' actions result from how their software or hardware is designed, responsibility for autonomous weapons should lie with the developers who create them. To the extent that their actions are enabled or constrained by civilian and military officials in their chain of command, those officials should share responsibility for the actions of autonomous weapons. (2013, 204)

This approach has been accepted in the past when considering the development of other weapons (Weeramantry 1985; Glerup and Horst 2014; Miller 2018; Khosrow-Pour D.B.A. 2021). However, it does not address the cases of AWS, insofar it remains oblivious to the possibility that, once deployed, AWS may develop behaviour independently from the intentions of their designers, developers, and users. It is worth clarifying here that the moral responsibility for this unintended behaviour is not akin to that for systems' failure, as the latter can be discharged referring to negligence, that is, the responsibility for some form of perturbation that the human agents could have and should have considered and prevented but did not. When considering the unintended behaviour of AWS, we are dealing with behaviour that may emerge also as a *correct* consequence of the technical features of the systems and/or their interactions with the surrounding environment, but which could not be anticipated by the human agents, as mentioned in Chapter 1.

At the same time, the focus on the group may lead to ascribing moral responsibilities, and blame/praise, to organisations rather than to individuals. As Taylor suggests:

[a] number of distinct groups might be identified as potential loci of responsibility: the government, the military, and the developers of LAWS. Of these, I suggest that most progress in closing the responsibility gap can be made by exploring the possibility of ascribing responsibility to the organisations that design and develop LAWS. (2020, 327)

This approach offers a limited solution, insofar as it is problematic to consider these groups to be *intentional* agents, as Corlett argued above. To overcome this issue, one may consider ascribing moral responsibility to individuals as representatives of groups or institutions (Champagne and Tonkens 2015; Galliott 2017). In this case, responsibility is attributed according to the role—and the obligations and duties attached to it—that individuals have, and not necessarily because of their intentions or the factual connection between cause and effect (Hin-Yan Liu 2016). Champagne and Tonkens (2015), for instance, propose that persons of sufficiently high military or civilian standing hold responsibility for the incorrect deployment of AWS simply by virtue of their office. By occupying high office, they argue, the occupant "willingly agrees" to the conditions of that office and, thereby, could in principle be made responsible for the use of AWS despite the latter's unpredictable nature.

In a sense, these proposals accord with the arrangements and values of a democratic system, that the privileges of office are attended by certain responsibilities, as well as by liability for consequences, not all of which are foreseeable or causally connected to the occupant of that office (Haddon 2020). However, a closer analysis reveals that, at best, they ascribe moral responsibility *nominally*—responsible human agents are identified because of their role more than because of their intentions, decisions, and actions. This risks creating a scapegoat, and thus makes moral responsibility less meaningful. The reader may recall the quotation from the Nuremberg trials, stressing the need to establish meaningful individual responsibility for any wrongs committed during the waging of war.

The next section analyses in more detail approaches focusing on the distribution of moral responsibility across the chain of command.

4.1. Moral Responsibility for AWS: Distributing Moral Responsibility along the Chain of Command

Approaches that support the idea of distributing moral responsibility for the actions of AWS along the chain of command rest on two assumptions, according to which responsibility should be shared

(i) proportionally to the decision-making power and the access to information characterising the different positions along the chain of command; and

(ii) according to the chain of command (within a military organisation) because of the authority that higher-ranking personnel have over the autonomy of lower-ranking personnel.

Assumption (i) resonates with the argument proposed by Walzer, according to which we have higher standards for commanders not just because of the dangerous instruments they have at their disposal but also because they have "access to all available information and also to the means of generating more information" (1977, 317). Assumption (ii) follows from (i) and is reflected in the rules of engagement (ROE) that defence organisations define prior to engaging in operations. ROE are established by cascading levels of hierarchy, with each level imposing a greater degree of constraint—and specificity—upon its subordinate level(s). This means that, at each level down the chain there is a (diminishing) degree of discretion and higher degree of specificity of possible decision/action, and it can be difficult or impossible to act contrary to the decisions made by superiors. Thus, personnel higher in the chain of command take responsibility for the actions performed by lower-ranking personnel in executing their orders, given the latter lack the autonomy to do differently than they are ordered. This view is shared by a good deal of Just War theorising (Walzer 1977), albeit not all (McMahan 2006). As Walzer puts it: "We regard soldiers under orders as men whose acts are not entirely their own and whose liability for what they do is somehow diminished" (1977, 309). This diminished autonomy, and thus diminished responsibility, is reflected in the doctrine of command responsibility, according to which superiors are held accountable for their subordinates by the principle of omission (e.g., a failure to prevent or intervene), when at least in principle they have sufficient control over their subordinate to prevent, or at least intervene to limit, immoral behaviour.

When considering AWS, assumptions (i) and (ii) are both problematic, insofar as they disregard the characteristics of AWS and the pragmatic and conceptual issues that may follow from their deployment. Let us consider assumption (i) first. It conflates the breadth of the decision-making power and of the importance of the information accessed with the level of granularity of the supporting information. For example, in some circumstances the risks and advantages of deploying AWS in a specific theatre of operation may be clearer to lower-ranking personnel, especially those familiar with the technology and the context, than to the higher-ranking personnel who may lack information specific to the technology or the context of deployment. And while information about risks and benefits of a specific operation may be conveyed upward in the chain of command, it is likely that in a context when AWS are deployed routinely and massively, the granularity of this information will decrease as the information is passed up (Payne 2021, 110–12). This may lead to personnel in higher command being held morally responsible for the actions of AWS, while not having access to information of sufficient granularity for this responsibility to be attributed in a justified and fair way. It should be noted that this is a pragmatic problem. It is logically possible that military institutions may put in place adequate processes to overcome it. In this case assumption (i) remains valid. But until a system is established to ensure that decision-makers have prompt access to sufficiently granular information about the benefits and risks of using AWS, the claim that the moral responsibility for the actions of AWS can be distributed with the level of command is unsound.

Assumption (ii) poses conceptual problems concerning control and autonomy. Let us start with control. The moral responsibilities of commanders for the actions of their subordinates rest on three conditions:

(1) the existence of a superior-subordinate relationship where the superior has effective control over the subordinate; (2) the requisite mental element, generally requiring that the superior knew or had reason to know (or should have known) of the subordinates' crimes; and (3) the failure to control, prevent or punish the commission of the offences. (Jain 2016, 310)

Establishing control, whether effective, appropriate, or meaningful of AWS has proven to be problematic (Ekelhof 2019a). This may be in part due to the circumstances of deployment of AWS (consider, for example, the cases of humans on- or post-loop) or to the characteristics of the context of deployment, of the type of AWS, or a combination of thereof. The predictability problem means that once deployed, AWS may perform unforeseen, unintended, and possibly, unwanted actions. This implies that officers deploying AWS may have no way to foresee unwanted outcomes and prevent them, making it hard to meet conditions (1) and (2), as specified by Jain. Condition (3) holds commanders responsible for the misconduct of their subordinates, but this is only insofar as they can control and prevent them from misbehaviour. In the case of AWS, their lack of predictability makes the attribution of this moral responsibility unjustified.

4.2. Moral Responsibility for AWS: The Distributed Faultless Moral Responsibility Approach

Some of the limitations of the approaches I have described so far can be overcome when considering distributed faultless moral responsibility.

According to this framework, commanders would be responsible for the actions of AWS to roughly the same extent as they are now, as they have similar powers to constrain the autonomy of AWS as they have over human soldiers. . . . The exact apportionment of blame between commanders and developers can be determined only by the extent to which they contribute to an AWS' wrongful actions through their actions or inactions. (Schulzke 2013, 216)

This approach allows for allocating moral responsibility to all the individuals participating in (and determining) the design, development, and deployment of AWS; but it is difficult to ascribe moral praise or blame in a justified way if we follow it. This is due to two reasons: lack of transparency and lack of intentionality.

The reverse engineering process necessary to identify the network of agents that shaped (causally) the behaviour of the AWS may be hindered by lack of transparency of the system itself, or by the limited transparency and traceability of the information about the system and the decision-making process underpinning its use (Tsamados et al. 2021a). The risk is concrete. It may arise as a consequence of the countless interactions among the many agents that shape the actions of AWS, and which may be difficult to reconstruct with sufficient detail to understand the factors that determined the behaviour of the system. State actors may also decide not to share relevant information. In 2010, the UN special rapporteur on extrajudicial, summary, or arbitrary executions stressed in a report on targeted killing that states may decide not to use "the procedural and other safeguards in place to ensure that killings are lawful and justified, and the accountability mechanisms that ensure wrongful killings are investigated, prosecuted and punished" (Alston 2010, 10). As Verdiesen, Santoni de Sio, and Dignum stress in their commentary on the report:

[t]he reason for this accountability vacuum is that the international community cannot verify the legality of the killing, nor confirm the authenticity of the intelligence used in the targeting process or ensure that the unlawful targeted killing results in impunity. (2021, 145)

At the same time, as we saw in section 2 of this chapter, reconstructing the causal chain of decisions and actions that led to a specific behaviour of AWS may not be sufficient to identify any intention for that behaviour to have occurred. Indeed, this approach aims at identifying *faultless* moral responsibility and does not aim to ascribe praise or blame. Thus, it sheds only limited light on the AWS responsibility gap, insofar as praise and blame are necessary to reward morally sound uses of these systems and punish, and remedy for, the moral evil that these uses of may cause.

The time has come to consider meaningful moral responsibility for the use of AWS, and the moral gambit that I propose to discharge this responsibility.

5. Meaningful Moral Responsibility and the Moral Gambit

When considering AWS, moral responsibility refers to responsibility for any actions leading to the destructive damage (whether lethal or not) that these systems may cause. One of the conditions for the use of these systems to be morally acceptable is that it must be possible to ascribe responsibility for any such damage in a justified way. This occurs when the outcomes of the systems reflect the intentionality of the human agents, and when there is a causal connection between the decisions/actions of the human agents and the AWS outcomes. It is also important that moral responsibility is attributed fairly, such that the agents involved have sufficient information and understanding of the context in which they operate to consider all possible alternatives before making any decision. Justified and fair moral responsibility is only attributed when all the four conditions specified in section 2 are met. Also, the agents have to be able to accept this moral responsibility as an element of their actions and decisions and have to be able to take the praise or blame that follows as an assessment of their moral character. *Meaningful* moral responsibility can be ascribed when all these criteria are met.

Discharging meaningful moral responsibility for the actions of AWS is a necessary, preliminary condition to their use, because it is the kind of responsibility that shows a minimum due care (Strawson 1962) for the receiver of the actions of AWS. In this sense Sparrow is correct in stressing that "the least we owe our enemies is allowing that their lives are of sufficient worth that someone should accept responsibility for their deaths" (2007a, 67). Meaningful moral responsibility enables backward-looking responsibility, as it fosters accountability. It may also enable forward-looking responsibility, insofar as the prospect of the praise and blame linked to a given decision/action should facilitate morally sound choices and careful conduct.

When considering AWS, the unpredictable nature of these systems limits meaningful moral responsibility. It would be unjustified and unfair to *ascribe* moral responsibility to human agents for (all possible) unpredictable actions that AWS may perform. This is because these actions are not intended by the human agent (moral responsibility would not be justified) and no information would allow the agent to foresee the totality of possible actions that an AWS may perform once deployed, to identify and prevent those unwanted (moral responsibility would not be fair). All one may ask is for the designers, developers, and deployers of AWS to *take* meaningful moral responsibility for the intended actions while being aware of the risk that unpredicted outcomes may occur and *taking* moral responsibility for the unpredictable effects that may follow the decision to deploy AWS. Let me clarify this point.

In taking this responsibility, the human agents make a *moral gambit*: they design/develop/use an AWS, being fully aware of the risks that it may perform some unpredicted actions. To limit these risks (and optimise the chances for a successful gambit) they, and the relevant defence institutions, must act at their best to establish all possible measures to constrain the moral evil (and harness the moral good) that unpredicted behaviour may cause. The human agents remain aware that independently of all these efforts, it will not be feasible to predict all possible actions of AWS and their effects in the context of deployment.[2] Nonetheless, if they decide to proceed with the design/development/use of these systems, then they make a moral gambit and decide to be morally responsible for the unforeseen AWS outcomes and their effects.

When considering AWS, all the human agents intentionally participating in the design, development, and use of these systems take this moral gambit. Personnel who decide on the deployment of these systems take (or not) the gambit only insofar as they have a choice as to use or not use AWS. Hence the responsibility gap for the actions of AWS may be closed so long as we can identify intentionality, causality with respect to the decision to use of AWS, full awareness about unpredictability of these systems, and willingness to take the moral gambit. As we will see in the next section, for this to be possible, an infrastructure for accessing relevant information, to ensure traceability of processes, and non-lethal outcomes needs to be established. Before moving to the next section, three clarifications are necessary to outline the boundaries of the gambit.

The moral gambit does not concern the decision to participate in an operation whose outcomes are relatively unpredictable. Rather it is about—having decided to participate—the *voluntary acceptance* of moral responsibility for the range of possible outcomes that may follow the use of AWS, whether foreseen or not. In this sense the moral gambit cannot be mandated; it must be taken in a voluntary way, and the responsibility that comes with it is *taken,* not attributed. The moral gambit is about taking ex ante responsibility for whatever may happen in that specific operation, hoping that unintended and unwanted outcomes never occur, but also being willing to be held responsible if they do.

---

[2] As mentioned in Chapter 1, the unpredictability is constrained within predetermined boundaries such as payload or range.

This takes us to the second point to clarify: the approach underpinning the gambit. The moral gambit places a heavier burden on the human agents to take the gambit and the moral responsibility that comes with it. This is why those who accept the gambit need to be fully informed and aware of risks and consequences of their choices, and institutions have a fundamental duty to support these individuals. This heavier weight offers a way to overcome the responsibility gap, while ensuring the moral responsibility remains meaningful and fair.

The third clarification concerns the permissibility of the moral gambit. In the context of national defence, when considering non-lethal AWS, the moral gambit may be acceptable. This is not the case when considering LAWS. In this case, the moral gambit would be a gambit taken on the lives of others. This i moraslly unacceptable.

The impermissibility of taking the moral gambit on the lives of others breaks down in two ways depending on the recipient of acts of war. Pertaining to non-combatants, the moral gambit is straightforwardly ruled out by the principle of distinction. Distinction provides non-combatants with immunity from attack in all times of war. Given the predictability problem, there is no certainty that LAWS would respect distinction (Blanchard and Taddeo 2022b) (see Chapter 8) and it would morally unacceptable to take a gambit as to whether the system would be able or not to respect this principle.

The impermissibility of the moral gambit on the lives of combatants is more complex. It is worth stressing that, in this case, the problem is with the mode of killing and not with killing *per se* (Blanchard and Taddeo 2022c). One could argue that combatants forgo their right not to be killed and that therefore it makes no difference if they are killed by humans or by LAWS (Walzer 1977, 42; Meisels 2018, 11–29) or whether it is possible to ascribe moral responsibility for their deaths. Hence, the permissibility of the gambit is a moot point. Whether combatants waive their right to life is a controversial point (Kamm 2004; McMahan 2011; Bazargan 2014), but let us accept it for the sake of argument.

The issue here is not *who* is liable to be killed, but *how* it is acceptable to kill those liable to be killed. Combatants may waive their right to life, but they do so under expectation that an attack on their life will comply with the principle of military necessity and respect the principle of the moral equality of combatants. Now, it is conceivable, though implausible, that LAWS may be used against human combatants in a way that complies with military necessity. It is conceivable because the Just War doctrine of supreme emergency justifies the use of any means if the stakes are high enough to warrant them—that is, survival (Walzer 1977, 251–68). It is implausible both because such circumstances seldom exist and because the operational advantages imputed to LAWS— like decision-making speed— exceed the capabilities of humans, thereby lowering the threshold for the necessity of using them against human combatants.

At the same time, the use of LAWS also breach the principle of moral equality of combatants. Under this principle, combatants are said to enter into a martial contract entailing mutual belligerency rules, whereby waiving of the right not to be killed is attended by a system of norms. As Skerker et al. write:

> [s]ervice members can be modelled as ceding claim-rights against being targeted with lethal violence by enemy combatants according to the terms of military norms optimizing a balance between maximising the one military's interests whilst minimizing suffering to their enemy. (Skerker, Purves, and Jenkins 2020, 202)

One of these norms is the expectation that combatants will not be targeted in a way that is either wanton or cavalier. This implies that unintended killing (whether of combatants or not) is at the very least morally problematic, because even if combatants waive their right not to be killed, they do so under the assumption that their lives will be taken intentionally, not as a result of an unintended behaviour or a lost moral gambit. Personnel choosing to deploy a LAWS would *take a gambit* that it will act according to the intended use, while being aware that there is a chance that a LAWS may identify, select, and engage an unintended target. This gambit, I argue, is incompatible with the expectations imposed by the principles of necessity and moral equality of combatants. This is why, insofar as LAWS are unpredictable with regard to the identification, selection, and engagement of human targets, it is not possible to ascribe meaningful moral responsibilities for their actions, and for this reason their use is morally impermissible.

6. Discharging Meaningful Moral Responsibility for the Actions of Non-lethal AWS

Much of the debate on AWS centres on LAWS, as lethal uses of AWS involve loss of life and thus severe ethical problems. Non-lethal AWS, by contrast, have attracted less attention. I believe that this gap is problematic. Thee ethical risks posed by non-lethal AWS may be less severe for impact and magnitude than those posed by LAWS, but these remain nevertheless serious. Non-lethal AWS can cause significant damage, including bodily harm, disproportionate destruction to property, infringements of liberty, and breaches of the principle of distinction. These ethical risks arise, for example, if non-lethal AWS are deployed "when the use of graduated force is required and deadly force is the exception" (Heyns 2016a, 5). Thus, attributing moral responsibility for the use of non-lethal AWS is crucial. The proposed moral gambit offers a new and much-needed solution in this respect.

The moral gambit cannot be imposed upon a human agent, nor can it come with the role. For it to be acceptable, the agent must make it willingly. When considering the use of *non-lethal AWS*, there are a number of procedures that can be established to support the decision to take, or not to take, the moral gambit. In what follows, I offer eight recommendations that tech providers in the defence domain and defence organisations should follow to support their members in this sense. The recommendations are listed in logical order.

1. Procure AWS whose underlying AI systems are *interpretable* and not just explainable. Lack of predictability of AI systems is partly a function of their lack of transparency. Explanations of a black-box model may offer imprecise representation of the original model (Rudin 2019); for this reason explainability offers limited solutions to the problems posed by lack of predictability. Better outcomes may be reached by procuring interpretable models, i.e. a model that is "constrained in model form so that it is either useful to someone, or obeys structural knowledge of the domain, such as monotonicity, causality, structural (generative) constraints, additivity or physical constraints that come from domain knowledge" (Rudin 2019, 206).

2. Assess predictability. Providers and defence institutions should assess which technical and operational features of AWS encroach upon the predictability of their ability to identify, select, and engage targets and work to improve them to limit unpredictable outcome.

3. High level of knowledge and understanding for the decision-makers. Those deciding about the use of non-lethal AWS should have a high level of technical understanding of these systems and of the theatre of operation, so as to be able to identify their strategic and tactical potential, how to deploy them best, and their possible points of failure. A high level of technical understanding underpins the choice to use non-lethal AWS insofar as it enables the decision-makers to

consider the risks of unpredictable behaviour properly and their implications for their moral gambit.

4. Traceability of processes. Information about the technical specification of a non-lethal weapons systems, their cycle of development, and their mode of procurement should be transparent to decision-makers. In the same way, any relevant information about the system that may advance the understanding of that decision-maker should be relayed to the personnel promptly and accurately.

5. Justification of uses. The decision to use or not use non-lethal AWS should always follow a risk-benefit analysis and be justified according to the principle of necessity. A further consideration should also be made about the safety of military personnel deployed in the specific theatre.

6. Ensuring non-lethal effects. Measures must be put in place to minimise the risks of lethal outcomes from the use of non-lethal AWS. These may not differ too much from similar measures taken when using conventional weapons and could include a range of approaches such as necessity and proportionality calculations, assessment of the theatre of operation, and user interface design features like a remote switch button.

7. Redressing and remedy. A process to identify mistakes and unwanted outcomes, to assess their impact and costs, and to define redressing and remedy measures should be established. Redressing and remedy measures will not override the moral praise or blame attributed to human agents, but should be used to discharge the accountability that a defence organisation has with respect to the decisions made by its personnel.

8. Auditing. Ethics-based auditing of both the non-lethal AWS and of the processes for their acquisition and deployment should be established (Mökander and Floridi 2021b), with the aim of facilitating accountability as well as to identify

possible points of failure and address them promptly, so as to improve the decision-making and the redressing processes.

7. Conclusion

In this chapter, I have argued that, for the use of AWS to be ethically acceptable, it is crucial to attribute *meaningful* moral responsibility to human agents. The attribution of this kind of moral responsibility rests on strong requirements, which are justified because of the nature of the damage that AWS may cause (whether lethal or not). As we saw in the previous pages, these requirements cannot be met when considering the case of LAWS, and, thus, the deployment of these systems is morally unacceptable, as it would undermine the morality of war. Instead, meaningful moral responsibility can be taken by means of the moral gambit for the actions of nonlethal AWS, provided that defence institutions establish necessary processes to support human agents willing to take the gambit in understanding the functioning of AWS, the benefits and risks linked to their uses, and the consequences of the moral gambit. I hope that the eight recommendations provided in this chapter will help technology providers and defence institutions to this end.

8

Just War Theory and the Permissibility of Autonomous Weapons Systems

1. Introduction

The time has arrived to consider the question of the moral permissibility of AWS. Discussions on the moral and legal permissibility of AWS date back to 2012, with the publication of the US DoD Directive on autonomy in weapon systems (US Department of Defense 2012). The DoD published an updated version of the directive in 2023. In the eleven years separating the two directives, the debate on AWS has shifted from being a speculative discussion on possible, but yet to happen, uses of these weapon systems to an urgent debate to settle the conditions, if any, for the legitimate use of these weapons, as records of their use began to emerge. Indeed, in March 2021 the UN released a document reporting the first official use of AWS on the Libyan front (Choudhury et al. 2021). It stresses that these systems "were programmed to attack targets without requiring data connectivity between the operator and the munition: in effect, a true "fire, forget and find" capability" (Choudhury et al. 2021, 17). If the use of AWS on the Libyan front could be considered the first (and perhaps isolated) case, the wide deployment of these weapons on both sides of the Ukraine war (US Department of Defense 2022a; Tiwari 2023; Knight 2022) broke the taboo on the use of AWS and has made even more compelling the need to define regulations for this type of weapon.

The Ethics of Artificial Intelligence in Defence. Mariarosaria Taddeo, Oxford University Press.
© Oxford University Press 2024. DOI: 10.1093/oso/9780197745441.003.0008

In this scenario, it is problematic that the debate on the permissibility of the use AWS has not progressed as quickly and efficiently as the development of these capabilities, and hence that the first reported uses of these weapons in Libya and Ukraine happened in a regulatory vacuum. This vacuum is the result of several factors, among which is the polarisation of the debate on the permissibility of AWS. As the adoption of AWS grows, it is imperative to move past this stalled debate and find a shared set of normative assumptions and desirable outcomes that may guide us in understanding limit cases of the use of AWS, that is, uses whose outcomes are morally unacceptable and are thus forbidden, while establishing norms for the other cases that may be deemed ethically and legally permissible. Note that this is not to say that a more balanced discussion on AWS is necessary because, given that AWS are *already* in use on the battleground, questions about their permissibility are now irrelevant. The point is rather that, given the current and foreseeable growing adoption of AWS, the question of their permissibility needs to be settled with a wider consensus that would enable to enforce any resulting measure—whether a ban or regulation of use. To this end, answers to this question need to start from premises that can be shared by both sides of the debate.

Just War Theory is central to this debate and its polarisation, with some maintaining the AWS can be used while respecting the principles Just War Theory and the IHL norms that it underpins (Arkin 2009; K. Anderson, Reisner, and Waxman 2014), while other argue that the use of AWS breaches these principles and key foundational values of our societies and is therefore impermissible (Marchant et al. 2011; Grut 2013; Foy 2014; Roff 2015; van den Boogaard 2015; Beard 2018; Davison 2018; Winter 2018, 2020). This polarisation results from radically different interpretations of Just War Theory, one stressing its consequentialist aspect and the other focusing on its deontological one.

The way in which the principle of distinction is interpreted by those pro and contra AWS is a paradigmatic example of the differences in the interpretation of Just War Theory and their consequences. The principle of distinction enshrines the protection of non-combatants in war waging, mandating that in planning and executing war operations, non-combatants should never be the intended target. When applied to the use of AWS, some maintain that these weapon systems can respect the principle of distinction better than human agents. Two reasons are often mentioned to support this view. First, as AWS have no sense of self-preservation, they can act in a more conservative manner when target identification is uncertain. If the principle of 'first do no harm' is encoded in the system, as Arkin writes, AWS would take a higher risk on themselves to protect non-combatants (2018, 3). Second, AWS may reduce instances of target misidentification, as unlike humans they do not try to fit or distort information into familiar patterns (Arkin 2009).

This view rests on a simplistic account of ethical reasoning and of the consequentialist element of Just War Theory. It reduces the former to a mere set of rules, disregarding any reflective process that includes awareness and recognition of the context and of the others involved; and discounts from the latter any reflection on the impact that actions may have on the receiver. Such a simplistic approach to ethical reasoning is coupled to an optimistic assessment of the capabilities, predictability, and robustness of AI systems, far in excess of current development of this technology. For example, Arkin claims that as long as the principles of Just War Theory are encoded correctly, AWS will be able to meet them. As we saw in Chapter 1, AI systems remain vulnerable to attacks and have limited robustness and predictability, which undermines the idea that they will necessarily behave as the designers, developers, and users intended, for example always respecting the principles of Just War Theory. The combination of the simplistic view and optimistic assessment trivialises ethical and technical problems, missing nuance, possible trade-offs, and risks. In so doing, it contributes to a polarisation of the debate on AWS.

An equally polarised view follows from analyses that leverage the deontological element of Just War Theory to argue against the use of AWS. According to this view, Just War Theory is not wholly consequentialist (Moseley 2011) and is centred around the duty to recognise and respect humanity. Following this view, respect as the ability to recognise the unique human value of the other in the battlefield is a crucial element in upholding the morality of war and one that AWS cannot meet, because recognising the dignity of a human being and their intrinsic value is different from identifying a target. For example, Asaro argues that legitimate killing requires that combatants understand each other's decision to enter warfare and respect it by recognising the significant value of the self-determining nature of the opponent, and being able to reflect on and endorse the reasons that justify their killing (Asaro 2020). In a similar vein, Sparrow builds on Nagel's view (1972) and argues for the need of direct relationship between the combatant and the target (Sparrow 2016b), because such a relation allows appreciating and recognising a potential target, as an autonomous individual, which is essential to respect the principle of distinction. AWS will breach the principle of distinction insofar as they sever this relation:

> the more advocates of robotic weapons laud their capacity to make complex decisions without input from a human operator, the more difficult it is to believe that AWS connect the killer and the killed directly enough to sustain the interpersonal relationship that Nagel argues is essential to the principle of distinction. (Sparrow 2016b, 108)

I agree with the view, but it also leads to a polarisation of the debate, as it does not engage with legitimate reasons offered in support of the use of AWS: military superiority, deterrence value, effectiveness, safety of the combatants, and the possibility of concluding wars more quickly and hence reduce risks to non-combatants. The polarisation of the debate favours the regulatory gap.

To develop effective regulation, it is important to establish a shared set of assumptions, values, and principles that would enable a constructive debate on what uses of AWS are morally acceptable, if any. Identifying this shared ground is the goal of this chapter. As a domain-specific ethical theory, Just War Theory can provide this ground; the question is how to interpret it to avoid the polarisation I have just described.

Here I offer an interpretation of Just War Theory that balances its consequentialist elements with the respect for humans qua humans and their dignity. To do so, I will rely on the analysis of Just War Theory and AWS provided elsewhere by Blanchard and myself (Blanchard and Taddeo 2022a, 2022b, 2022c). I begin by focusing on *jus ad bellum* principles of last resort and proportionality, to show that when applied to the case of AWS,[1] these principles do not offer meaningful guidance on the permissibility of these weapons. I then move onto the *jus in bello* principle of necessity, arguing that it has been neglected in the relevant literature mistakenly and that this principle is crucial in assessing the permissible uses of AWS. I then move to the principle of distinction, to offer an argument that balancing military necessity with the deontological and consequentialist elements of Just War Theory is feasible, opening the way to consider limit cases for the use of AWS while offering guidance for the regulation of other uses that may be deemed permissible.

---

[1] A clarification is necessary before beginning with the analysis. In the rest of this chapter, I shall refer to AWS as defined in Chapter 6; hence the focus will be on autonomous artificial agents endowed with learning capabilities that allow them to adapt their behaviour to the conditions of deployment and which are used to apply force, whether lethal or not. I will use "AWS" to refer both lethal and non-lethal autonomous weapon systems. I shall specify when this is not the case and the analysis is focusing on a specific subset of AWS, e.g., LAWS or non-lethal AWS.

2. *Jus ad bellum* and AWS

Both sides of the debate on the permissibility of AWS refer to *jus ad bellum*—the part of Just War Theory that defines principles for, and the conditions under which, a state may resort to war or to the use of force—in particular to the principles of last resort and proportionality. Both lines of argument rest on commonsensical interpretations of these principles. As Blanchard and I have shown elsewhere (2022c), arguments based on *jus ad bellum* do not offer a solid ground to motivate the use or the ban of AWS. The goal of this section is to clear the ground of theses in order to focus the debate on more relevant aspects of Just War Theory.

Let me begin by focusing on the principle of last resort, according to which war is to be the least preferred option.[2] The principle mandates that political leaders assess all available means other than war to meet a given threat, and to opt for non-violent means sufficient for doing so (Coverdale 2004, 258–59). It is worth remembering that the principle of last resort does not mean that political leaders should turn to war when no other possible course of action is available. Interpreting last resort in this way would mean that war can never be justified since it is always possible to say that not every alternative has been tried. Rather, the principle of last resort "requires a considered judgement about whether some imagined alternative has a good chance of avoiding war. It does not require that every idea actually be pursued to the end of the line" (Allen, quoted in Coverdale 2004, 259). Those who object to the use of AWS maintain that AWS would reduce the economic, political, and human costs of war, making the choice to go to war more convenient and incentivising political leaders to declare war

---

[2] Much of the opposition to AWS on the basis of last resort parallels concerns that have been raised over the use of drones in warfare (see Strawser 2013). For sake of clarity, here I maintain a focus here on AWS.

instead of pursuing alternative means to resolve a conflict, thus violating the principle of last resort (Asaro 2008).

However, last resort prescribes that the leader needs to consider alternative, available, and effective means first. Thus, no matter whether AWS provide an incentive in going to war, if other means have not been exhausted properly, the decision to resort to war remains unjust. Thus, the objection that AWS would lead to breaching the principle of last resort is not inherent to AWS but rather linked to the willingness of political leaders to abide by this principle. More sophisticated work on technology and Just War Theory should consider not only the ethical problem that technology may pose, but also how it informs judgements concerning proportionality and last resort. In that context, the above objection to AWS has some value, insofar as new technological developments require awareness of unpredicted and unwanted effects, like escalation (Allenby 2013).

Precision is often cited by both sides of the debate on AWS as another element that impacts *ad bellum* assessment, particularly proportionality. Both sides of the debate refer to *ad bellum* proportionality and to the impact of AWS on the likelihood that political leaders who can use AWS may decide to go to war. On the one side, AWS are deemed to reduce the time span of hostilities and thus reduce the potential damage and harms to non-combatants— and hence foster respect of *ad bellum* proportionality. On the other side, opponents to the use of AWS agree on the precision of AWS as a factor that will reduce the harms caused by war, but maintain that in doing so AWS lower the barriers to warfare (Enemark 2011; Brunstetter and Braun 2013; Asaro 2008; Roff 2015) and that this may lead to an increase in the number of wars (Abney 2013, 340).

In both cases, analyses rest on a superficial understanding of proportionality, because they confuse proportionality, a contextual assessment, with precision, an objective property of weapons (Braun and Brunstetter 2013).[3] *Ad bellum* proportionality prescribes that

---

[3] The confusion generated by overlooking such factors has a long pedigree and became clera in the initial deployment of drones in combat settings and the argument,

when political leaders decide whether to enter war, the destruction expected to be caused by the war must be proportionate to the good that is expected to result (Hurka 2005, 35). It requires a global outlook, and estimates must include expected deaths and structural and economic damages. Precision of a weapon is crucial in the *in bello* proportionality assessment but is irrelevant in the *ad bellum* proportionality assessment, which has to do with the assessment of overall purposes sought in the pursuit of a war, the tactics, the types of costs envisioned, and so forth.

This assessment is already difficult and indeterminate enough when considering foreseeable wars. The difficulty increases when assessment needs to factor in the effects of a given technology. As Sechser et al. write: "Extrapolating from current technological trends is problematic, both because technologies often do not live up to their promise, and because technologies often have countervailing or condition effects that can temper their negative consequences" (Sechser, Narang, and Talmadge 2019, 728). Attempts to make this assessment for wars that are not yet to happen and unforeseeable are speculative, and this is why they do not support arguments about the likelihood of incidence of war vis-à-vis AWS.

This is why *jus ad bellum* does not provide a solid ground on which to settle issues about the permissibility of AWS. Arguments pro and contra AWS that rely on last resort and *ad bellum* proportionality refer to political attitudes of leaders more than to the moral prescription of these principles, and thus may shed more light on decision-making processes in the international arena than on the moral permissibility of AWS *per se*. How political leaders make

---

made by their proponents, that, once viewed on an "historical trend line", drones can be understood as generating more proportionate wars (K. Anderson 2012, 383–84). However, often this "historical trend line" took World war II as its baseline, and given that the Allied forces intentionally targeted civilian populations, it seemed not much praise to model drones against a "historic nadir for warfare " (Braun and Brunstetter 2013, 309; see also Shue 2008).

decisions in the context of war and how technological capabilities, and the related narrative, may impact these decisions are important topics of analysis but do not pertain to the moral permissibility of AWS. The principles of *jus in bello* provide a more solid ground on which to base this analysis. I shall turn to them in the next two sections.

3. The Principle of Necessity

The principles of necessity, distinction, and proportionality are central in *jus in bello*—the part of Just War Theory that sets the principles for contact in war. The principle of necessity allows for measures that are necessary to accomplish a legitimate military objective. It involves both a permission and a constraint. The permission states that such measures are justified if and only if they are permissible (i.e., they respect the principles of distinction and proportionality) and also *necessary* to achieve the given goal. As Lackey formulated it, "[t]he principle [of necessity] does not say that whatever is necessary is permissible, but that everything permissible must be necessary" (Lackey, quoted in Ohlin and May 2016, 77). The constraint states that unnecessary measures are unjustified and must be avoided (Matsumoto 2020). The constraint therefore introduces the requirement of minimal force, which poses a moral obligation for combatants to use the minimal necessary force to achieve a legitimate military objective (McMahan and McKim 1993, 516–17; Lango 2010, 482).

When considering AWS, the principle of necessity best corresponds to the principle of justified use of AI that I outlined in Chapter 1, which cautions against overuse, thereby generating risks, or underuse, thereby generating opportunity costs (Taddeo, McNeish, et al. 2021). In this sense, necessity, along with the principles of distinction and proportionality, offers essential guidance to assess the permissibility of AWS, as it outlines the risks of combatants being harmed and balances this with the need to achieve military objectives. Nonetheless, necessity has been overlooked in the relevant literature (Grut 2013; Wagner 2014), with some scholars maintaining that it does not apply to AWS or does not shed significant light on the permissibility of these weapon systems. For example, Foy rejects the principle of necessity as unhelpful in considering the permissible use of AWS, claiming that "while the use of AWS engages the principles of military necessity and unnecessary suffering, these principles are engaged in a different way than distinction and proportionality" (2014, 54). This is because, according to Foy's analysis, the principle of necessity is relatively unaffected by autonomy in weapons systems. This objection is misplaced, however, because the principle of necessity and the requirement of minimal force imply some level of control over the effects of a weapon, whether this is autonomous or not (Blanchard and Taddeo 2022a). Thus, they are no less applicable to AWS than they are to other means and methods of warfare. At the same time, control of effects is a core issue when considering the permissibility of AWS (Tsamados and Taddeo 2023), and considering it within the normative boundaries of the principle of necessity will clarify what levels of control are required for AWS to be considered permissible.

To understand the requirement of minimal force it is important to consider how the principle of necessity differs from, and interacts with, the principles of proportionality and distinction. Often the principle of necessity is confused for the principle of proportionality (Chengeta 2016, 131). However, the two prescribe different calculations. An act of war can be proportionate because its costs are tolerable relative to its benefits, but it can also be unnecessary because those benefits could have been achieved by less costly means (Hurka 2008, 128). Conversely, an act of war can be necessary insofar as it is the least destructive—or perhaps the only— means to attain a given objective, while failing to be proportionate, because the costs would not be tolerable relative to the benefits.

At the same time, the principles of necessity and proportionality address different actors. The latter refers to the unintended but foreseeable harms done to non-combatants, while the former considers the harms done to combatants (McMahan 2009, 19–23). Thus proportionality does not make necessity redundant. This is why objections that consider the principle of necessity redundant in assessing the permissibility of AWS are mistaken. Consider, for example, Schmitt and Thurner's point according to which permissible uses of AWS can be defined on the basis of the principles of proportionality and distinction, without the need to consider necessity:

> as to prohibitions based on use, the requirement that military objectives yield some military advantage would make any separate condition for military necessity redundant. With regard to situations raising proportionality issues, any strike lacking military advantage but causing harm to civilians or civilian objects would violate the rule . . . the law of armed conflict already prohibits attacks on those who have surrendered or are otherwise hors de combat. Taking these observations together, the result is that military necessity has little or no independent valence when assessing the legality of autonomous weapon systems or their use. (2012, 258–59)

Indeed, given the distribution of risk and liability in war, distinction and proportionality remain crucial (Blanchard and Taddeo 2022b). However, the assumption underpinning this objection is misconceived because distinction, proportionality, and necessity address different risks of war. The three principles inform one another and provide sufficient guidance for combatants only if they are considered together. Necessity is of particular relevance when considering AWS, as by focusing on harms posed to combatants it addresses questions related to the respect and rights of combatants, which are central to the debate on AWS and which are not addressed by the principles of proportionality and distinction.

The question here concerns the obligations that enemy combatants owe to each other. If necessity is discounted, it follows that an enemy combatant (who is not surrendering or *hors de combat*) is susceptible to any amount of force. This is precisely what the requirement of minimal force is set to avoid, as it introduces a moral obligation for belligerents to use the minimum necessary force to realise a legitimate military objective (McMahan and McKim 1993; Lango 2010). The requirement is defined in the preamble to the 1868 St. Petersburg Declaration:

> that the progress of civilization should have the effect of alleviating as much as possible the calamities of war;
> That the only legitimate object which States should endeavour to accomplish during war is to weaken the military forces of the enemy;
> That for this purpose it is sufficient to *disable* the greatest possible number of men;
> *That this object would be exceeded by the employment of arms which uselessly aggravate the sufferings of disabled men, or render their deaths inevitable;*
> That the employment of such arms would, therefore, be contrary to the law of humanity. (International Committee of the Red Cross 2020, emphasis added)

Anything beyond what is minimally required is impermissible. The requirement encompasses both lethal and non-lethal uses of force and entails a moral obligation to use non-lethal means to incapacitate an enemy combatant where lethal means would be unnecessary (in light of a legitimate military objective) (Kaurin 2010). This is why Childress argues that the primary objective of a belligerent "is not to kill or even to injure any particular person, but to incapacitate or restrain him" (cited in Lango 2010, 484). With the requirement of minimal force, Just War Theory defines a requirement for non-lethal means. Therefore, discounting necessity in relation to

AWS means discounting the question of whether it is possible to use AWS at the threshold below lethality.

3.1. The Principle of Necessity and AWS

Lethal and non-lethal uses of a weapon can be hard to distinguish. As a 1972 report on non-lethal weapons by the US National Science Foundation put it:

> all weapons . . . create some primary or secondary risk of death or permanent injury. The probable seriousness of their effects (their lethality) depends on a number of factors, not all of which are determined by their design. Weapons not intended to kill or create permanent injury, if used with an degree of regularity, would undoubtedly cause some deaths because of the physiological differences among those against whom they employed, physical malfunctioning, improper utilization, and other circumstances. (National Science Foundation, cited in Davison 2009, 1).

The complexity of combat environments worsens the indeterminacy of effects. Thus, distinguishing lethal or non-lethal use of a weapon rests on the assessment of its intended purpose of use (Ramsey 2002).

To respect the principle of necessity and meet the requirement of minimal force, the intention of the combatant needs to be supported by contextually bound judgements on the application of the necessity principle in the specific context (Ohlin and May 2016, 86). Here the point is that military necessity is in part governed by the principle of success in a given set of circumstances: if an action does not increase the likelihood of attaining a military objective, then that action is unnecessary to meet the objective. Success has two elements: luck and ability. Luck depends on whether circumstances outside the control of the human combatants are conducive to achieving the goal. Ability is the capacity of the human combatant to achieve a goal taking into consideration the circumstances. Ability therefore requires a recognition of one's capability to achieve that goal, along with an ability to factor that recognition into calculations about success (Ohlin and May 2016). As mentioned in the previous section, when considering AWS, difficulties emerge when undertaking such an assessment because of the limited predictability of these systems.

The predictability problem (as outlined in Chapter 1) has two relevant implications here. The first one concerns the intention of the human combatant to use AWS to apply lethal or non-lethal force. As we saw in Chapter 7, the limited predictability of the autonomous system decouples the intent of the human combatant from the outcome of the use of AWS. A human combatant may use AWS in a way that was intended to be non-lethal, but the AWS may act in such a way as to result in lethal effects. This undermines the possibility of using AWS according to the requirement of minimal force. The second reason has implications for the ability of the human combatant to determine the chances of success when employing AWS in specific contexts. Success is dependent on ability, and ability requires recognition of one's capacity to achieve the legitimate military goal and a capacity to factor that recognition into calculations about success. However, this assessment is undermined by the uncertainty of the outcomes of deploying AWS in any given context. Until these unknowns are addressed, it will be problematic to use AWS in compliance with the requirement of minimal force. This is especially the case in circumstances where the requirement calls for the use of non-lethal force that requires stricter control on the effects of a military action.

It is important to note that the challenges related to meeting the requirement of minimal force refer to concrete settings—that is, the possibility for the human combatant to develop situational judgement—rather than to the principle of necessity itself. In principle, these challenges may be met if appropriate measures are put in place to mitigate the risks posed by the predictability problem. These could be, for example, setting and enforcing a risk threshold for an acceptable level of unpredictability of AWS (see Chapter 4 and the next section). Treating decision-makers, human combatants, and AWS as agents of a hybrid system rather than agents acting along a chain of command may also help to this end (Taddeo et al. 2022). In this way, training protocols for human agents to foster appropriate levels of trust in the technology, maintain situational awareness, and deal with possible unpredicted outcomes could be designed to improve human control of AWS. Technical standards could also be defined that specify suitable human-machine interfaces and identify contexts of deployment of non-lethal AWS, for example submarine environments, where the risks related to the predictability problem would have a limited impact.

I shall turn to this point in the following section. First let me recall the arguments against the relevancy of the principle of necessity in the debate on AWS. I hope that the analysis presented here has convinced the reader of the importance of this principle. I shall add two more points to support this view. The first concerns the scope of the debate, which has so far been centred on LAWS and has disregarded non-lethal uses of AWS. This is problematic because non-lethal uses of AWS pose ethical problems that are not addressed in discussions on LAWS. The debate on LAWS focuses on killing; thus it centres on the permissibility of taking life and overlooks key problems concerning intentionality of use, domination, coercion, and autonomy that emerge clearly when focusing on non-lethal uses. Non-lethal AWS are more likely to be permissible than LAWS, which is why it is crucial to include them in the debate and address the related ethical risks. The second point speaks to the goal of this chapter, that is, to identify a shared ground between opposite positions of the debate on AWS to avoid the continued polarisation of the discussion. The identification of trade-offs between military necessity and minimal force for non-lethal uses of AWS could inform the debate on LAWS as well. To put it differently, if non-lethal uses of AWS cannot be used in compliance with the requirement of minimum force, it is unlikely that *any* use of AWS, whether lethal or not, may be considered morally permissible. But if acceptable trade-offs to justify non-lethal uses can be identified, they may also shed light on the conditions of permissibility of the use of LAWS.

4. Distinction, Double Effect, and Due Care

The principle of distinction requires parties to an armed conflict distinguish between military objects and civilian objects, and direct their actions only towards the former. The distinction is underpinned by liability and non-liability to attack. Combatants are liable to attack, for they have abrogated their right not to be attacked. Non-combatant immunity is absolute in Just War Theory (Walzer 1977, 151). Nonetheless, the principle of distinction has numerous interpretations (Bica 1998; Kasher 2007). Minimally, it states that combatants should not target intentionally non-combatants. However, the principle of distinction "does not make it illegal for civilians to die in wartime" *per se* (Orend 2019, 112), insofar as distinction permits the unintentional but foreseeable harming of non-combatants if that harm is proportionate to the goals the attack is intended to achieve. This is the doctrine of "double effect". The doctrine conveys the moral intuition that

> it is permissible to cause a harm as a side effect (or "double effect") of bringing about a good result even though it would not be permissible to cause such a harm as a means to bringing about the same good end. (McIntyre 2004)

Walzer outlined the four conditions of the doctrine of double effect in war, all of which must be met for harm to non-combatants to be permissible. These are:

1. "The act is good in itself or at least indifferent, which means . . . that it is a legitimate act of war.
2. The direct effect is morally acceptable—the destruction of military supplies, for example, or the killing of enemy soldiers.
3. The intention of the actor is good, that is, he aims only at the acceptable effect; the evil effect is not one of his ends, nor is it a means to his ends.
4. The good effect is sufficiently good to compensate for allowing the evil effect; it must be justifiable under . . . [the] proportionality rule" (Walzer 1977, 153).

Blanchard and I (Blanchard and Taddeo 2022b) have argued that these conditions give a minimalist account of the principle of distinction and that this account does not forbid the use of AWS. It is possible for a human combatant to employ AWS with good intention, aiming for an acceptable effect. It can be argued that the unpredictably of AWS implies that it was foreseeable that non-combatants would be harmed unintentionally, and if that good effect (i.e., the military objective) is "sufficiently good to compensate for allowing the evil effect" (Walzer 1977, 153) under the proportionality rule, then such a harm is permissible.

However, the minimalist account fails to consider the obligation of due care entailed by the principle of distinction (Walzer 1977, 151–59). This obligation is crucial when considering AWS (Blanchard and Taddeo 2022b), as it defines the risks and obligations of combatants in using (lethal) force. Due care states that combatants are obliged to accept greater risks to themselves to ensure that they hit only the right target in order to diminish the risks to non-combatants (Orend 2001, 12–13).[4] The nature of due care has not been given full articulation in Just War Theory, but it is widely held to be a central tenet of *jus in bello* conduct. As McMahan writes:

---

[4] Due care has been codified in the Law of Armed Conflict as precautions in attack (Ministry of Defence 2011, 81–88).

> [t]he dominant view within the just war tradition . . . is that when combatants must choose between imposing a certain risk on civilians as a side effect of their action and accepting an even greater risk to themselves, they must, up to a certain point, accept the greater risk. (2010, 344)

In a war, combatants are liable to harm because of their capacity to injure. It follows that if a combatant is liable to harm through intentional attack, then liability should also encompass liability to harm from the side effects of military action. Accordingly, it is part of the role of a combatant to accept greater risk (within the limits defined by the principle of necessity and the requirement of minimal force) and to avoid imposing them on those who are non-liable (Margalit and Walzer 2009). Due care affirms the good intention highlighted by the doctrine of double effect through the demonstration of restraint in warfare (Orend 2001, 13). It requires that the foreseeable harm caused by a military action is reduced as far as possible. Thus, it makes the doctrine of double effect and the principle of distinction more restrictive. When considering due care, the principle of distinction is satisfied if

> the intention of the actor is good, that is, he aims narrowly at the acceptable effect; the evil effect is not one of his ends, nor is it a means to his ends, and, aware of the evil involved, he seeks to minimize it, accepting costs to himself. (Walzer 1977, 155)

Thus, respecting the principle of distinction depends on whether risks of harm can be distributed appropriately between combatants and non-combatants, and also on the particular threshold at which combatants cease to be obliged to shoulder additional risk (Walzer 1977, 157).[5]

---

[5] By *justified* military objective is meant that the objective conforms with the other principles of *jus in bello* conduct, such as the principle of necessity and proportionality.

Meeting both conditions requires a situational assessment that accounts for "the nature of the target, the urgency of the moment, the available technology and so on" (Walzer 1977, 156). When considering AWS, the issue is that such an assessment is unfeasible given the predictability problem and the highly dynamic conditions of contemporary war. Indeed, the predictability problem motivates this ICRC recommendation:

> [*u*]*npredictable* autonomous weapon systems should be expressly ruled out, notably because of their indiscriminate effects. This would best be achieved with a prohibition on autonomous weapon systems that are designed or used in a manner such that their effects cannot be sufficiently understood, predicted and explained. (International Committee of the Red Cross 2021, 2, emphasis added)

However, this position assumes a naive dichotomy between predictable and unpredictable AWS, give the reference to "*unpredictable* autonomous weapon systems" and risks leading to a two-tiered approach to regulation, which fails to capture the more nuanced nature of the predictability problem and its wide ramifications. If limited to the technical aspects of AWS, this distinction is correct in principle. There are models of AWS such as offline models that, considered in isolation, do not necessarily present an issue for predictability. However, as shown in Chapter 1, autonomous and learning systems, whether or not predictable by design, may exhibit unforeseen behaviour when deployed. The limited predictability of AWS is not contingent on their current state of development or conditions of deployment; rather all AI systems are unpredictable to a certain degree because of their autonomy, learning capabilities, and limited robustness. This is why the unpredictability of an AI system increases with its level of sophistication and with the complexities of adversarial conditions. As noted by the UNIDIR:

[t]he more complex the operating environment to which a system is deployed, the more likely it is that the system will encounter inputs for which it was not specifically trained or tested or will display new behaviours that have not been previously observed or validated. (Holland Michel 2020b, 7)

If we consider these aspects, it is evident that a clear-cut distinction between predictable and unpredictable AWS should be abandoned, and instead the focus should be on *levels of predictability*, related risks, and the specification of which steps, if any, of the process of exerting force may be within the remit of AWS given their level of predictability, the risks it poses, and the obligation of due care. I turn to these points in the next section.

4.1. AWS, Distinction, and Due Care

At the tactical level of mission execution a series of tasks precedes the decision to engage the target and exert force, including finding, fixing, and tracking the target. Due care requires combatants to determine the liability of the target on the basis of their situational assessment. The delegation of this task to AWS poses a high risk of breaching the obligation of due care, because of the limits of AWS due to the predictability problem and the complexity of the contexts it will need to assess.

When considering AWS and their capability to identify legitimate targets, the literature refers to cases where an AWS *sees* objects or individuals and identifies them targets. However, in contemporary warfare, it is unlikely that liability to be attacked can be determinable on the basis of purely visual markers, for example, wearing a uniform. Visual markers worked at a time when warfare took place between soldiers operating at some distance from civilian centres (Nurick 1945). Today, asymmetric urban insurgencies involve non-uniformed combatants and demand a much greater use of situational awareness to determine the liability of an agent to attack, because the assessment of this liability depends on the behaviour of the target, and often behavioural differences are highly nuanced. For example, civilians using a weapon for the protection of their family, and not to gain any advantage in the conflict, are not liable to attack.

Categories such as combatant versus non-combatant and soldier versus civilian introduce near-ideal situations, which do not hold in reality, because the status of an agent changes with their behaviour. This is why the recommendation to use AWS only to target "objects that are military objectives by nature" (International Committee of the Red Cross 2021, 2) is problematic, because whether an actor or object is a military objective can be determined only within a specific context. For instance, a combatant who has either surrendered or been incapacitated through injury becomes *hors de combat* and non-liable to attack. One might reply that there are artefacts such as tanks which are objectively means of war, constituting essentially military targets irrespective of context of use. But this is not always the case (Blanchard and Taddeo 2022b). Consider, for example,

> the so-called "Highway of Death" in first Gulf War (Mueller 1995). On the night of 26th of February 1991 coalition aircraft attacked and destroyed a column of hundreds of Iraqi manned military vehicles. At the time the controversy of the incident rested on the dispute over whether Iraqi forces were surrendering or retreating to regroup. If the former, then the attacks launched by the coalition forces were potentially in violation of the Geneva Convention (Hersh 2000). The existence of this dispute testifies to the fact that the notion of objects which are military objectives by nature is misconceived. (Blanchard and Taddeo 2022b, 19)

Thus, determining the liability or non-liability of attacking within a specific context requires an assessment based on a nuanced, situational awareness of that context. This does not have to be an assessment of the target qua human, and it does not require an appreciation of the dignity and a recognition of the respect due to another human being, but it must be an assessment of the target to be harmed that is at least as accurate and certain as would have been undertaken by a human combatant operating in the same circumstances. In principle, it is not impossible for AWS to run such an assessment, but it is unlikely that they will reach such a level of precision and certainty in the foreseeable future.

The limits of AWS in identifying legitimate targets outlined in this section motivate the ICRC recommendation to ban these weapon systems because they bring risks of harm for those affected by armed conflict. However, I believe that distinction and due care need to be considered along with necessity and proportionality, and also along with a consideration of military necessity and the operational advantages that AWS could offer. Taken together, these factors call for a more nuanced position.

All weapon systems, whether autonomous or not, pose risks to those affected by armed conflict. The question is whether these risks are morally (and legally) acceptable. To address this question, it is crucial to determine risk thresholds above which the risks introduced by the limited predictability of AWS are not acceptable, as well as to define risk-mitigating measures, like testing, training, software validation, and cybersecurity measures. These points are gaining consensus among practitioners and policymakers. For example, the UN CCW GGE (2019) states: "Risk assessments and mitigation measures should be part of the design, development, testing and deployment of emerging technologies in any weapons systems." I agree that the focus on risk and risks thresholds is a viable approach to avoid the polarisation of the debate on the permissibility of AWS and may foster a more fruitful debate to close the regulatory gap and identify limit cases for the use of these weapons.

However, there is a cautionary note to sound here, because the focus on risk management may become a shortcut to evade the reflection on Just War Theory and ethics of war that should ground any regulatory approach to the use of AWS. It is crucial that the debate on the permissibility of AWS is not reduced to a simple risk-management issue, thereby overlooking the normative factors that should drive the distribution of risks and the definition of acceptable risk thresholds mentioned in this chapter.

The principles of necessity and distinction both rest on the question of risk allocation and how to resolve the trade-off between force protection and minimising the harm of war on both combatants and non-combatants (McMahan 2010, 343). Answers to these questions are not value neutral; they are informed by moral and political choices (Perry 1995; Garland 2003). Thus, the definition of such threshold and risk-mitigating measures requires a public debate, a multi-stakeholder forum, and a normative ground to inform discussion. Just War Theory offers such a ground; what is needed is a model for the distribution of risks to assess whether these would be of a magnitude and impact and could be distributed in such a way, that is consistent with the principles of this theory. Without such a solid normative ground, risk-assessment approaches to AWS would not settle the debate on the permissibility of AWS.

5. Conclusion

The goal of this chapter has been to offer an interpretation of Just War Theory that can provide a shared ground among the different sides of the debate on the permissibility of AWS, in the hope of overcoming the current polarisation and addressing the regulatory gap about the use of these weapons. The analysis of the principles of necessity and distinction in this chapter shows that when focusing on risk thresholds, it is possible to identify limit cases to the use of AWS without relying exceedingly on the consequentialist or deontological element of Just War Theory, and while also factoring in military necessity.

This analysis leads to three key conclusions. First, respecting the principles of necessity and distinction requires appropriate situational judgement and awareness to assess combat scenarios, related risks, and make justifiable strategic and tactical decisions. The question as to whether AWS could be left in charge of such an assessment while respecting these principles is not worth asking. On the one side, the idea that these systems may develop an appropriate level of situational awareness to deal with the intricacies of contemporary war scenarios is unrealistic. On the other side, limited predictability and robustness indicate that AWS may exhibit behaviour different from the one expected, posing severe risks that they may breach the principles that they should respect.

Second, AWS may bring important tactical and strategic advantages that may justify their use, but given the analysis proposed here, this use is to be envisaged as complementary to, rather than in lieu of, human combatants. AWS should be framed as part of a hybrid team where they collaborate with humans. When framed in this way, questions about the permissibility of AWS can be settled addressing issues as to the context in which it acceptable to deploy human-machine teams in war, what type of tasks can be delegated to AWS, what level of autonomy they should have, what type of training is required for human combatants working with AWS, how to design interfaces so as to enhance transparency and control over AWS, and how to identify, assess, and mitigate related risks and define risk thresholds and standards to assess the predictability of AWS (Tsamados and Taddeo 2023).

Third, the idea of using AWS in full autonomous mode remains morally unacceptable, as this use would pose too severe a risk of breaching the principles of Just War Theory. Ultimately, any use of AWS needs to be regulated accounting for military necessity but, most important, according to what is acceptable while upholding the values and rights that underpin them.

Epilogue

Heaven doth with us as we with torches do
Not light them for themselves; for if our virtues
Did not go forth of us, 'twere all alike
As if we had them not.

—William Shakespeare, *Measure for Measure*,
Act 1, Scene 1

I am writing this epilogue nearly two years after the war in Ukraine began and a few months following the conflict in Gaza. Against the backdrop of these conflicts, I have found myself wondering whether a book like this one is useful at all. An ethics of war may be justified and necessary, yet it risks irrelevance if state actors ignore its prescriptions. Consider that, for example, both sides in the Ukrainian war have potentially fully autonomous weapon systems while the ethical debate on their permissibility is still ongoing, with many stakeholders calling for a prohibition of their use. The reported uses of AI systems to target humans in Gaza without appropriate human oversight and the consequent reported breaches of the principles of proportionality and necessity also seem to point to the irrelevancy of an ethics of AI in defence.

After a time of disheartenment, the harsh unfolding of these two conflicts has nudged me towards at least attempting to make a difference. As Walzer put it, "The restraint of war is the beginning of peace" (2006, 335). In today's digital era, the beginning of peace means curtailing the capabilities of digital technologies, especially AI, used in waging war.

There is relation of mutual influence between the way conflicts are waged and the societies waging them. As Clausewitz remarked, more than an art or a science, wars are a social activity. And much like other social activities, conflicts mirror the values of societies while relying on their technological and scientific developments. In turn, the principles we use to regulate warfare play a crucial role in shaping societies. Think about the design, deployment, and regulation of weapons of mass destruction (WMDs). During World War II, WMDs were made possible by scientific breakthroughs in nuclear physics, which was a central area of research in the years leading up to the war. However, the catastrophic violence that was unleashed on the cities of Hiroshima and Nagasaki led to a global consensus that was hitherto unseen, which shaped the post-war world's aversion to the use of such weapons. The Cold War that followed and the nuclear treaties that ended it defined the modes in which nuclear technologies and WMDs can be used, drawing a line between conflicts and atrocities. In doing so, treaties and regulations for the use of WMDs helped to shape contemporary societies as they rejected the belligerent rhetoric of the early twentieth century, striving instead for peace and stability.

Things are not different with today's digital societies and the wars they wage. On the one side, the use of AI in defence holds great potential to improve the working of defence organisations, increase defence capabilities, and make military operations safer, more effective, and more efficient. On the other side, we know that if left unregulated the use of AI in defence can have serious negative ramifications in terms of stability, as outlined in Chapter 4, of individual and group rights, of breaches of the principles of Just War Theory, as I argued in Chapters 3–8. Consider for example, the case of open-source intelligence in the ongoing Russo-Ukrainian conflict. The widespread use of smartphones among citizens has enabled military personnel to exploit intelligence gathered by the civilian population and shared on social media to obtain approximate location estimates for enemy combatants. This has raised significant concerns about the extent of military surveillance of civil society and the risks of involving the civilian population in military operations. The question is whether this is a trade-off of our values and rights that we are willing to accept.

An ethical framework about the use of AI in defence needs to be firm about the ethical risks and the limit cases, but it also has to be able to identify the good potential of AI for defence and offer guidelines to leverage it in accordance to the values underpinning our societies. This balancing act is quite complex. It demands substantial investment of time and resources and requires a join effort from scholars, tech developers, policymakers, and defence practitioners. Scholars, particularly ethicists, play a key role in identifying ethical values, principles, a value theory, and even guidelines to implement their frameworks. Given the current geopolitical scenario and the pace of adoption of AI in defence, this work is quite urgent. However, the great burden lies with state actors and the defence establishment. They need to consider the ethical challenges of AI and adopt adequate, robust, and independently developed ethical frameworks to meet these challenges genuinely. This implies accepting that, and seeking to develop and adopt, an ethical governance of AI in defence needs be as comprehensive and profound as the changes that this technology poses. In the end this is the only way to ensure that AI will work as a technology for stability, and possibly peace, rather than a mere means of war.

References

Abney, Keith. 2013. "Autonomous Robots and The Future of Just War Theory". In *Routledge Handbook of Ethics and War: Just War Theory in the Twentieth-First Century*, edited by Fritz Allhoff, Nicholas G. Evans, and Adam Henschke, 338–51. London: Routledge.

"Acalvio Autonomous Deception". 2019. *Acalvio.* https://www.acalvio.com/. Accessed July 2024.

Adeney, Douglas, and John Weckert. 1997. *Computer and Information Ethics.* Westport, CT: Praeger.

Akhgar, Babak, and Simeon Yates. 2013. *Strategic Intelligence Management: National Security Imperatives and Information and Communications Technologies.* Waltham, MA: Elsevier/Butterworth-Heinmann.

Alexy, Robert. 2002. *A Theory of Constitutional Rights.* New York: Oxford University Press.

Aljunied, Syed Mohammed Ad'ha. 2020. "The Securitization of Cyberspace Governance in Singapore". *Asian Security* 16 (3): 343–62. https://doi.org/10.1080/14799855.2019.1687444.

Allenby, Braden. 2013. "Emerging Technologies and Just War Theory". In *Routledge Handbook of Ethics and War: Just War Theory in the Twenty-First Century*, edited by Fritz Allhoff, Nicholas G. Evans, and Adam Henschke, 289–300. London: Routledge.

Alshammari, Majed, and Andrew Simpson. 2017. "Towards a Principled Approach for Engineering Privacy by Design". In *Privacy Technologies and Policy*, edited by Erich Schweighofer, Herbert Leitold, Andreas Mitrakas, and Kai Rannenberg, 10518:161–77. Cham: Springer International Publishing. https://doi.org/10.1007/978-3-319-67280-9_9.

Alston, Philip. 2010. "Report of the Special Rapporteur on Extrajudicial, Summary or Arbitrary Executions, Philip Alston: Addendum—Study on Targeted Killings (A/HRC/14/24/Add.6)—Russian Federation". May 28, 2010. *ReliefWeb.* https://reliefweb.int/report/russian-federation/report-special-rapporteur-extrajudicial-summary-or-arbitrary-executions.

Amoroso, Daniele, and Guglielmo Tamburrini. 2020. "Autonomous Weapons Systems and Meaningful Human Control: Ethical and Legal Issues". *Current Robotics Reports* 1 (4): 187–94. https://doi.org/10.1007/s43154-020-00024-3.

Anderson, Andrew, Jonathan Dodge, Amrita Sadarangani, Zoe Juozapaitis, Evan Newman, Jed Irvine, Souti Chattopadhyay, Matthew Olson, Alan Fern, and Margaret Burnett. 2020. "Mental Models of Mere Mortals with Explanations of Reinforcement Learning". *ACM Transactions on Interactive Intelligent Systems* 10 (2): 1–37. https://doi.org/10.1145/3366485.

Anderson, David. 2016. *Report of the Bulk Powers Review*. London: Independent Reviewer of Terrorism Legislation. https://terrorismlegislationreviewer.inde pendent.gov.uk/wp-content/uploads/2016/08/Bulk-Powers-Review-final-rep ort.pdf.

Anderson, Kenneth. 2012. "Efficiency *in Bello* and *ad Bellum*: Making the Use of Force Too Easy?" In *Targeted Killings: Law and Morality in an Asymmetrical World*, edited by Claire Finkelstein, Jens David Ohlin, and Andrew Altman, 374–99. New York: Oxford University Press.

Anderson, Kenneth, Daniel Reisner, and Matthew C. Waxman. 2014. "Adapting the Law of Armed Conflict to Autonomous Weapon Systems". *International Law Studies* 90: 386–411.

Andras, Peter, Lukas Esterle, Michael Guckert, The Anh Han, Peter R. Lewis, Kristina Milanovic, Terry Payne, et al. 2018. "Trusting Intelligent Machines: Deepening Trust within Socio-technical Systems". *IEEE Technology and Society Magazine* 37 (4): 76–83. https://doi.org/10.1109/MTS.2018.2876107.

Aquin, Mathieu d', Pinelopi Troullinou, Noel E. O'Connor, Aindrias Cullen, Gráinne Faller, and Louise Holden. 2018. "Towards an 'Ethics by Design' Methodology for AI Research Projects". In *Proceedings of the 2018 AAAI/ACM Conference on AI, Ethics, and Society*, 54–59. New Orleans, LA: Association for Computing Machinery. https://doi.org/10.1145/3278721.3278765.

Arkin, Ronald. 2009. "Ethical Robots in Warfare". *IEEE Technology and Society Magazine* 28 (1): 30–33.

Arkin, Ronald. 2018. "Lethal Autonomous Systems and the Plight of the Non-combatant". In *The Political Economy of Robots*, edited by Ryan Kiggins, 317–26. Cham: Springer.

Arquilla. 1999. "Ethics and Information Warfare". In *Strategic Appraisal: The Changing Role of Information in Warfare*, edited by Zalmay Khalilzad and John Patrick White, 379–401. Santa Monica, CA: RAND.

Article36. 2018. "Shifting Definitions—the UK and Autonomous Weapons Systems July 2018". http://www.article36.org/wp-content/uploads/2018/07/Shifting-definitions-UK-and-autonomous-weapons-July-2018.pdf.

Asaro, Peter. 2008. "How Just Could a Robot War Be". In *Proceeding of the 2008 Confernece on Current Issues in Computing and Philosophy,* edited by Adam Briggle,Katinka Waelbers and Philip A. E. Brey, 50–64, Amsterdam, the Netherlands: IOS Press.

Asaro, Peter. 2012. "On Banning Autonomous Weapon Systems: Human Rights, Automation, and the Dehumanization of Lethal Decision-Making". *International Review of the Red Cross* 94 (886): 687–709. https://doi.org/10.1017/S1816383112000768.

Asaro, Peter. 2020. "Autonomous Weapons and the Ethics of Artificial Intelligence". In *Ethics of Artificial Intelligence*, edited by S. Matthew Liao, 212–36. New York: Oxford University Press.

Athalye, Anish, Logan Engstrom, Andrew Ilyas, and K. Kwok. 2018. "Synthesizing Robust Adversarial Examples". June 7, 2018. https://www.semanticscholar. org/paper/Synthesizing-Robust-Adversarial-Examples-Athalye-Engstrom/8dce99e33c6fceb3e79023f5894fdbe733c91e92.

Avishai, Margalit, and Michael Walzer. 2009. "Israel: Civilians & Combatants". *New York Review of Books*, May. https://www.nybooks.com/articles/2009/08/13/israel-civilians-combatants-an-exchange.

Ayling, Jacqui, and Adriane Chapman. 2022. "Putting AI Ethics to Work: Are the Tools Fit for Purpose?" *AI and Ethics* 2 (3): 405–29. https://doi.org/10.1007/s43 681-021-00084-x.

Babuta, Alexander, and Marion Oswald. 2020. "Data Analytics and Algorithms in Policing in England and Wales: Towards A New Policy Framework". Occasional paper. London: Royal United Services Institute for Defence Studies.

Babuta, Alexander, Marion Oswald, and Ardi Janjeva. 2020. "Artificial Intelligence and UK National Security: Policy Considerations". Occasional paper. London: Royal United Services Institute for Defence Studies.

Bazargan, Saba. 2014. "Killing Minimally Responsible Threats". *Ethics* 125 (1): 114–36. https://doi.org/10.1086/677023.

Beard, Jack M. 2018. "The Principle of Proportionality in an Era of High Technology". In *Complex Battlespaces: The Law of Armed Conflict and the Dynamics of Modern Warfare*, edited by Christopher M. Ford and Winston S. Williams, 261–70. New York: Oxford University Press.

Bebber, Robert. 2018. "There Is No Such Thing as Cyber Deterrence. Please Stop". *Cypher Brief*, April 1, 2018. https://www.thecipherbrief.com/column_article/no-thing-cyber-deterrence-please-stop.

"BehavioSec: Continuous Authentication through Behavioral Biometrics". 2019. BehavioSec. https://www.behaviosec.com/.

Bekele, Esube, Wallace E. Lawson, Zachary Horne, and Sangeet Khemlani. 2018. "Human-Level Explanatory Biases for Person Re-identification". *HRI 2018: Explainable Robotic Systems* 2: 1–2.

Bekele, Esube, Cody Narber, and Wallace Lawson. 2017. "Multi-attribute Residual Network (MAResNet) for Soft-Biometrics Recognition in Surveillance Scenarios". In *2017 12th IEEE International Conference on Automatic Face & Gesture Recognition (FG 2017)*, 386–93. Washington, DC: IEEE. https://doi.org/10.1109/FG.2017.55.

Bendiek, Annegret, and Tobias Metzger. 2015. "Deterrence Theory in the Cyber-Century: Lessons from a State-of-the-Art Literature Review". In Lecture Notes in Informatics (LNI), 553–70. Bonn: Gesellschaft für Informatik.

Bentham, Jeremy. 1789. *An Introduction to the Principles of Morals and Legislation*. Garden City, NY: Doubleday.

Bergadano, F. 1991. "The Problem of Induction and Machine Learning". In *IJCAI'91: Proceedings of the 12th International Joint Conference on Artificial Intelligence*, Vol. 2, 1073–78. San Francisco: Morgan Kaufmann Publishers Inc.

Bernal, Paul. 2016. "Data Gathering, Surveillance and Human Rights: Recasting the Debate". *Journal of Cyber Policy* 1 (2): 243–64. https://doi.org/10.1080/23738871.2016.1228990.

Betz, D. J., and T. Stevens. 2013. "Analogical Reasoning and Cyber Security". *Security Dialogue* 44 (2): 147–64. https://doi.org/10.1177/0967010613478323.

Bica, Camillo C. 1998. "Interpreting Just War Theory's *Jus in Bello* Criterion of Discrimination". *Public Affairs Quarterly* 12 (2): 157–68.

Biggio, Battista, and Fabio Roli. 2018. "Wild Patterns: Ten Years after the Rise of Adversarial Machine Learning". *Pattern Recognition* 84 (December): 317–31. https://doi.org/10.1016/j.patcog.2018.07.02.

Biometrics and Surveillance Camera Commissioner. 2017. *National Surveillance Camera Strategy for England and Wales*. Whitehall: Biometrics and Surveillance Camera Commissioner.

Blanchard, Alexander, and Mariarosaria Taddeo. 2022a. "*Jus in Bello* Necessity, the Requirement of Minimal Force, and Autonomous Weapon Systems". *Journal of Military Ethics* 21 (3–4): 286–303. https://doi.org/10.1080/15027 570.2022.2157952.

Blanchard, Alexander, and Mariarosaria Taddeo. 2022b. "Predictability, Distinction & Due Care in the Use of Lethal Autonomous Weapons Systems". May 3, 2022. *SSRN Electronic Journal*. http://dx.doi.org/10.2139/ssrn.4099394.

Blanchard, Alexander, and Mariarosaria Taddeo. 2022c. "Autonomous Weapon Systems and *Jus ad Bellum*". *AI & Society*, March. https://doi.org/10.1007/s00 146-022-01425-y.

Blanchard, Alexander, and Mariarosaria Taddeo. 2023. "The Ethics of Artificial Intelligence for Intelligence Analysis: A Review of the Key Challenges with Recommendations". *Digital Society* 2 (1): 12. https://doi.org/10.1007/s44 206-023-00036-4.

Blanchard, Alexander, Chris Thomas, and Mariarosaria Taddeo. 2023. "Ethical Governance of Artificial Intelligence for Defence: Normative Tradeoffs for Principle to Practice Guidance". July 21, 2023. *SSRN Electronic Journal*. https://doi.org/10.2139/ssrn.4517701.

Boardman, Michael, and Fiona Butcher. 2019. "An Exploration of Maintaining Human Control in AI Enabled Systems and the Challenges of Achieving It". *STO-MP-IST-178*, 1–16.

Boca, Paul. 2014. *Formal Methods: State of the Art and New Directions*. London: Springer.

Bologna, Sandro, Alessandro Fasani, and Maurizio Martellini. 2013. "From Fortress to Resilience". In *Cyber Security: Deterrence and IT Protection for Critical Infrastructures*, edited by Maurizio Martellini, 53–56. Heidelberg: Springer.

Boogaard, Jeroen van den. 2015. "Proportionality and Autonomous Weapons Systems". *Journal of International Humanitarian Legal Studies* 6 (2): 247–83.

Boulanin, Vincent, Moa Peldán Carlsson, Netta Goussac, and Davison Davidson. 2020. "Limits on Autonomy in Weapon Systems: Identifying Practical Elements of Human Control". Stockholm International Peace Research Institute and the International Committee of the Red Cross. https://www.sipri.org/publications/ 2020/other-publications/limits-autonomy-weapon-systems-identifying-practi cal-elements-human-control-0.

Branscombe, Nyla R., Susan Owen, Teri A. Garstka, and Jason Coleman. 1996. "Rape and Accident Counterfactuals: Who Might Have Done Otherwise and Would It Have Changed the Outcome?". *Journal of Applied Social Psychology* 26 (12): 1042–67. https://doi.org/10.1111/j.1559-1816.1996.tb01124.x.

Braun, Megan, and Daniel R. Brunstetter. 2013. "Rethinking the Criterion for Assessing CIA-Targeted Killings: Drones, Proportionality and *Jus ad Vim*". *Journal of Military Ethics* 12 (4): 304–24.

Brent, Laura. 2019. "NATO's Role in Cyberspace". *NATO Review*, February 12, 2019. https://www.nato.int/docu/review/articles/2019/02/12/natos-role-in-cyberspace/index.html.

Brewster, Thomas. 2020. "Google Promised Not to Use Its AI in Weapons, So Why Is It Investing in Startups Straight out of 'Star Wars'?" Forbes, December 22, 2020. https://www.forbes.com/sites/thomasbrewster/2020/12/22/google-promised-not-to-use-its-ai-in-weapons-so-why-is-alphabet-investing-in-ai-satellite-startups-with-military-contracts/.

Brittain, Stephen. 2016. "Justifying the Teleological Methodology of the European Court of Justice: A Rebuttal". *Irish Jurist, new series* 55: 134–65.

Brodie, Bernard. 1978. "The Development of Nuclear Strategy". *International Security* 2 (4): 65–68.

Brundage, Miles, Shahar Avin, Jack Clark, Helen Toner, Peter Eckersley, Ben Garfinkel, Allan Dafoe, et al. 2018. "The Malicious Use of Artificial Intelligence: Forecasting, Prevention, and Mitigation". *arXiv:1802.07228 [Cs]*, February. http://arxiv.org/abs/1802.07228.

Brunstetter, Daniel, and Megan Braun. 2013. "From *Jus ad Bellum* to *Jus ad Vim*: Recalibrating Our Understanding of the Moral Use of Force". *Ethics & International Affairs* 27 (1): 87–106. https://doi.org/10.1017/S089267941 2000792.

Buchanan, Ben, and Andrew Imbrie. 2022. *The New Fire: War, Peace, and Democracy in the Age of AI*. Cambridge, MA: MIT Press.

Bunn, M. E. 2007. "Can Deterrence Be Tailored?" Strategic Forum, No. 225, January 2007. Washington, DC: Institute for National Strategic Studies, National Defense University.

Burgess, Matt. 2017. 'What Is the Petya Ransomware Spreading across Europe? WIRED Explains'. Wired, 2017, sec. Security. https://www.wired.com/story/petya-malware-ransomware-attack-outbreak-june-2017/.

Cambridge Consultants. 2019. "Use of AI in Online Content Moderation". Report prepared for Ofcom. https://www.ofcom.org.uk/__data/assets/pdf_file/0028/157249/cambridge-consultants-ai-content-moderation.pdf.

Cameron, David, and Great Britain Cabinet Office. 2010. *A Strong Britain in an Age of Uncertainty: The National Security Strategy*. London: TSO.

Campedelli, Gian Maria, Mihovil Bartulovic, and Kathleen M. Carley. 2021. "Learning Future Terrorist Targets through Temporal Meta-Graphs". *Scientific Reports* 11 (1): 1–15.

Castelfranchi, Cristiano, and Rino Falcone. 2003. "From Automaticity to Autonomy: The Frontier of Artificial Agents". In *Agent Autonomy*, edited by Henry Hexmoor, Cristiano Castelfranchi, and Rino Falcone, 103–36. Boston, MA: Springer US. https://doi.org/10.1007/978-1-4419-9198-0_6.

Cath, Corinne, Sandra Wachter, Brent Mittelstadt, Mariarosaria Taddeo, and Luciano Floridi. 2018. "Artificial Intelligence and the 'Good Society': The US, EU, and UK Approach". *Science and Engineering Ethics* 24 (2): 505–28.

CCW GGE. 2019. "Guiding Principles Affirmed by the Group of Governmental Experts on Emerging Technologies in the Area of Lethal Autonomous Weapons System (Annex III)". *Vol. CCW/MSP/2019/9*. Geneva: United Nations Office of Disarmament Affairs. https://www.ccdcoe.org/uploads/2020/02/UN-191

213_CCW-MSP-Final-report-Annex-III_Guiding-Principles-affirmed-by-GGE.pdf.

Champagne, Marc, and Ryan Tonkens. 2015. "Bridging the Responsibility Gap in Automated Warfare". *Philosophy & Technology* 28 (1): 125–37.

Chen, Jim Q. 2016. "Intelligent Targeting with Contextual Binding". In *2016 Future Technologies Conference (FTC)*, 1040–46. San Francisco, CA: IEEE. https://doi.org/10.1109/FTC.2016.7821732.

Chengeta, Thompson. 2016. "Measuring Autonomous Weapon Systems against International Humanitarian Law Rules". *Journal of Law & Cyber Warfare* 5 (1): 66–146.

China. 2018. "Convention on Certain Conventional Weapons: Position Paper Submitted by China". https://docs-library.unoda.org/Convention_on_Certai n_Conventional_Weapons_-_Group_of_Governmental_Experts_(2022)/CCW-GGE.1-2022-WP.6.pdf.

Chopra, Amit K., and Munindar P. Singh. 2018. "Sociotechnical Systems and Ethics in the Large". In *Proceedings of the 2018 AAAI/ACM Conference on AI, Ethics, and Society*, 48–53. New York: Association for Computing Machinery. https://doi.org/10.1145/3278721.3278740.

Choudhury, L., A. Aoun, D. Badawy, L. A. de Alburquerque, J. Marjane, and A. Wilkinson. 2021. "Letter [from] the Panel of Experts on Libya Established Pursuant to Resolution 1973 (2011) Addressed to the President of the Security Council". *s/2021/229. United Nations Security Council.*

Cicero, Marcus Tullius. 2008. *On Obligations.* Translated by Patrick G. Walsh. Oxford: Oxford University Press.

Cihon, Peter, Jonas Schuett, and Seth D. Baum. 2021. "Corporate Governance of Artificial Intelligence in the Public Interest". *Information* 12 (7): 275. https://doi.org/10.3390/info12070275.

Clark, David, and Susan Landau. 2011. "Untangling Attribution". *Harvard National Security Journal* 2011 (2): 25–40.

Clausewitz, Carl von. 2008. *On War.* Translated by James John Graham. Radford, VA: Wilder Publications.

Coldicutt, R., and C. Miller. 2019. "People, Power, and Technology: The Tech Workers' View". London: Doteveryone. https://doteveryone.org.uk/wp-cont ent/uploads/2019/04/PeoplePowerTech_Doteveryone_May2019.pdf.

Coleman, Stephen. 2015. "Possible Ethical Problems with Military Use of Non-lethal Weapons International Regulation of Emerging Military Technologies". *Case Western Reserve Journal of International Law* 47 (1): 185–200.

Collopy, Paul, Valerie Sitterle, and Jennifer Petrillo. 2020. "Validation Testing of Autonomous Learning Systems". Insight 23 (1): 48–51. https://doi.org/10.1002/inst.12285.

Conn, Ariel. 2016. "The Problem of Defining Autonomous Weapons". Future of Life Institute. November 30, 2016. https://futureoflife.org/2016/11/30/prob lem-defining-autonomous-weapons/.

Convention on Certain Conventional Weapons. 2018. "Report of the 2018 Session of the Group of Governmental Experts on Emerging Technologies in the Area of Lethal Autonomous Weapons Systems". *CCW/GGE.1/2018/3.* Geneva: United Nations Office for Disarmament Affairs. https://documents.un.org/doc/undoc/

gen/g18/323/29/pdf/g1832329.pdf?token=8FjuvBoEJj8vL89phx&fe=true. Accessed July 2024.

Corlett, J. Angelo. 2001. "Collective Moral Responsibility". *Journal of Social Philosophy* 32 (4): 573–84. https://doi.org/10.1111/0047-2786.00115.

Cornille, Chris. 2021. "AI Experts Needed to Lead 'Project Maven' Move within DOD". *Bloomberg Government* (blog), June 1, 2021. https://about.bgov.com/news/ai-experts-needed-to-lead-project-maven-move-within-dod/.

Coverdale, John F. 2004. "An Introduction to the Just War Tradition". *Pace International Law Review* 16 (2): 221–78.

Crosston, Matthew. 2011. "World Gone Cyber MAD: How 'Mutually Assured Debilitation' Is the Best Hope for Cyber Deterrence". *Strategic Studies Quarterly* 50 (1): 100–116.

Cummings, Mary, and Songpo Li. 2019. "HAL2019-02: Machine Learning Tools for Informing Transportation Technology and Policy". Humans and Autonomy Laboratory, Duke University. http://hal.pratt.duke.edu/sites/hal.pratt.duke.edu/files/u39/HAL2019_2%5B1920%5D-min.pdf.

"DarkLight Offers First of Its Kind Artificial Intelligence to Enhance Cybersecurity Defenses". 2017. Business Wire, July 26, 2017. https://www.businesswire.com/news/home/20170726005117/en/DarkLight-Offers-Kind-Artificial-Intelligence-Enhance-Cybersecurity.

Davies, Rachel, Jonathan Ives, and Michael Dunn. 2015. "A Systematic Review of Empirical Bioethics Methodologies". *BMC Medical Ethics* 16 (1): 15. https://doi.org/10.1186/s12910-015-0010-3.

Davison, Neil. 2009. *"Non-lethal" Weapons*. London: Palgrave Macmillan.

Davison, Neil. 2018. "A Legal Perspective: Autonomous Weapon Systems under International Humanitarian Law". UNODA Occasional Papers, no. 30, New York: United Nations.

"DeepLocker: How AI Can Power a Stealthy New Breed of Malware". 2018. *Security Intelligence* (blog), August 8, 2018. https://securityintelligence.com/deeplocker-how-ai-can-power-a-stealthy-new-breed-of-malware/.

Defense Innovation Board. 2017. "Defence Innovation Board Recommendations". https://media.defense.gov/2017/Dec/18/2001857962/-1/-1/0/2017-2566-148525_RECOMMENDATION%2012_(2017-09-19-01-45-51).PDF.

Defence Innovation Board. 2019. "AI Principles: Recommendations on the Ethical Use of Artificial Intelligence by the Department of Defence". https://media.defense.gov/2019/Oct/31/2002204458/-1/-1/0/DIB_AI_PRINCIPLES_PRIMARY_DOCUMENT.PDF. Accessed June 9, 2024.

Defense Technical Information Center. Department of Defense. 2013. "Joint Publication 2-0—Joint Intelligence". https://web.archive.org/web/20160613010839/http://www.dtic.mil/doctrine/new_pubs/jp2_0.pdf.

Dehousse, Renaud. 1998. *The European Court of Justice: The Politics of Judicial Integration*. New York: St. Martin's Press.

Department for Digital, Culture, Media & Sport. 2018. "Data Ethics Framework". https://www.gov.uk/government/publications/data-ethics-framework/data-ethics-framework.

Department of National Defence. 2018. "Autonomous Systems for Defence and Security: Trust and Barriers to Adoption. Innovation Network Opportunities".

Government of Canada. July 16, 2018. https://www.canada.ca/en/department-national-defence/programs/defence-ideas/current-opportunities/innovation-network-opportunities.html#ftn1.

Diller, Antoni. 1994. *Z: An Introduction to Formal Methods*. 2nd ed. New York: Wiley & Sons.

Ding, Wen, Sonwoo Kim, Daniel Xu, and Inki Kim. 2019. "Can Intelligent Agent Improve Human-Machine Team Performance under Cyberattacks?" In *2019 Intelligent Human Systems Integration*, edited by Waldemar Karwowski and Tareq Ahram, 725–30. Amsterdam, Netherlands: Springer. https://doi.org/10.1007/978-3-030-11051-2_110.

Docherty, Bonnie. 2014. "Shaking the Foundations: The Human Rights Implications of Killer Robots". *Human Rights Watch*. https://www.hrw.org/report/2014/05/12/shaking-foundations/human-rights-implications-killer-robots.

Docherty, Bonnie. 2020. "The Need for and Elements of a New Treaty on Fully Autonomous Weapons". Human Rights Watch. June 1, 2020. https://www.hrw.org/news/2020/06/01/need-and-elements-new-treaty-fully-autonomous-weapons.

DoD Responsible AI Working Council. 2022. "Responsible Artificial Intelligence Strategy and Implementation Pathway", Pittsburg, PA: Carnegie Mellon University, United States.

Donaldson, Thomas, and Lee E. Preston. 1995. "The Stakeholder Theory of the Corporation: Concepts, Evidence, and Implications". *Academy of Management Review* 20 (1): 65–91.

Doyle, Andy, Graham Katz, Kristen Summers, Chris Ackermann, Ilya Zavorin, Zunsik Lim, Sathappan Muthiah, et al. 2014. "Forecasting Significant Societal Events Using the Embers Streaming Predictive Analytics System". *Big Data* 2 (4): 185–95. https://doi.org/10.1089/big.2014.0046.

Dunnmon, Jared, Bryce Goodman, Peter Kirechu, Carol Smith, and Alexandrea Van Deusen. 2021. "Responsible AI Guidelines in Practice: Operationalizing DoD's Ethical Principles for AI". Defense Innovation Unit. https://assets.ctfassets.net/3nanhbfkr0pc/acoo1Fj5uungnGNPJ3QWy/3a1dafd64f22efcf8f27380aafae9789/2021_RAI_Report-v3.pdf.

Ehsan, Upol, and Mark O. Riedl. 2020. "Human-Centered Explainable AI: Towards a Reflective Sociotechnical Approach". In *HCI International 2020— Late Breaking Papers: Multimodality and Intelligence*, edited by Constantine Stephanidis, Masaaki Kurosu, Helmut Degen, and Lauren Reinerman-Jones, 449–66. Cham: Springer International Publishing. https://doi.org/10.1007/978-3-030-60117-1_33.

Eitel-Porter, Ray. 2021. "Beyond the Promise: Implementing Ethical AI". *AI and Ethics* 1 (1): 73–80. https://doi.org/10.1007/s43681-020-00011-6.

Ekelhof, Merel. 2019. "Moving beyond Semantics on Autonomous Weapons: Meaningful Human Control in Operation". *Global Policy* 10 (3): 343–48. https://doi.org/10.1111/1758-5899.12665.

Ekelhof, Merel, and Giacomo Persi Paoli. 2021. "The Human Element in Decisions about the Use of Force". March 31, 2020. UNIDIR. https://unidir.org/publication/the-human-element-in-decisions-about-the-use-of-force/.

Enemark, Christian. 2008. "'Non-lethal' Weapons and the Occupation of Iraq: Technology, Ethics and Law". *Cambridge Review of International Affairs* 21 (2): 199–215.

Enemark, Christian. 2011. "Drones over Pakistan: Secrecy, Ethics, and Counterinsurgency". *Asian Security* 7 (3): 218–37. https://doi.org/10.1080/14799855.2011.615082.

ENISA. 2020. "Artificial Intelligence Cybersecurity Challenges". https://www.enisa.europa.eu/publications/artificial-intelligence-cybersecurity-challenges.

Erwin, Sandra. 2017. "With Commercial Satellite Imagery, Computer Learns to Quickly Find Missile Sites in China". *SpaceNews*, October 19, 2017. https://spacenews.com/with-commercial-satellite-imagery-computer-learns-to-quickly-find-missile-sites-in-china/.

European Commission. 2021. AI Act Proposal. https://eur-lex.europa.eu/legal-content/EN/TXT/?uri=CELEX%3A52021PC0206.

European Union. 2014. "Cyber Defence in the EU: Preparing for Cyber Warfare? Think Tank". Brussels. http://www.europarl.europa.eu/thinktank/en/document.html?reference=EPRS_BRI(2014)542143.

European Union. 2015. "Cyber Diplomacy: EU Dialogue with Third Countries—Think Tank". Brussels. http://www.europarl.europa.eu/thinktank/en/document.html?reference=EPRS_BRI(2015)564374.

Evans, Michael. 2021. "Pentagon Uses AI to Predict Enemy Moves 'Days in Advance'". *The Times* (London), August 3, *2021, sec. World.* https://www.thetimes.co.uk/article/pentagon-uses-ai-to-predict-enemy-moves-days-in-advance-bql5q5s9p.

Eykholt, Kevin, Ivan Evtimov, Earlence Fernandes, Bo Li, Amir Rahmati, Chaowei Xiao, Atul Prakash, Tadayoshi Kohno, and Dawn Song. 2018. "Robust Physical-World Attacks on Deep Learning Visual Classification". In *2018 IEEE/CVF Conference on Computer Vision and Pattern Recognition*, 1625–34. Salt Lake City, UT: IEEE. https://doi.org/10.1109/CVPR.2018.00175.

Fabre, Cécile. 2009. "Guns, Food, and Liability to Attack in War". *Ethics* 120 (1): 36–63. https://doi.org/10.1086/649218.

Fang, Richard, Rohan Bindu, Akul Gupta, Qiusi Zhan, and Daniel Kang. 2024. "Teams of LLM agents can exploit zero-day vulnerabilities". *arXiv.* https://doi.org/10.48550/ARXIV.2406.01637.

Fazelpour, Sina, and Zachary C. Lipton. 2020. "Algorithmic Fairness from a Non-ideal Perspective". *In Proceedings of the AAAI/ACM Conference on AI, Ethics, and Society*, 57–63. New York: Association for Computing Machinery. https://doi.org/10.1145/3375627.3375828.

Federal Foreign Office. 2020. "German Commentary on Operationalizing All Eleven Guiding Principles at a National Level as Requested by the Chair of the 2020 Group of Governmental Experts on Emerging Technologies in the Area of Lethal Autonomous Weapons Systems within the Convention on Certain Conventional Weapons". https://documents.unoda.org/wp-content/uploads/2020/07/20200626-Germany.pdf.

Fennelly, Nial. 1997. "Legal Interpretation at the European Court of Justice". *Fordham International Law Journal,* 20 (3): 656–79.

Fischer, John Martin, and Mark Ravizza. 2000. *Responsibility and Control: A Theory of Moral Responsibility.* Cambridge: Cambridge University Press.

Floridi, L. 2014. *The Fourth Revolution: How the Infosphere Is Reshaping Human Reality.* Oxford: Oxford University Press.

Floridi, L., and M. Taddeo, eds. 2014. *The Ethics of Information Warfare.* New York: Springer.

Floridi, Luciano. 2006. "Information Ethics, Its Nature and Scope". *SIGCAS Computers and Society* 36 (3): 21–36. https://doi.org/10.1145/1195716.1195719.

Floridi, Luciano. 2008. "The Method of Levels of Abstraction". *Minds and Machines* 18 (3): 303–29. https://doi.org/10.1007/s11023-008-9113-7.

Floridi, Luciano. 2012. "Distributed Morality in an Information Society". *Science and Engineering Ethics* 19 (3): 727–43. https://doi.org/10.1007/s11948-012-9413-4.

Floridi, Luciano. 2013. *The Ethics of Information.* Oxford: Oxford University Press.

Floridi, Luciano. 2016. "Faultless Responsibility: On the Nature and Allocation of Moral Responsibility for Distributed Moral Actions". *Philosophical Transactions of the Royal Society A: Mathematical, Physical and Engineering Sciences* 374 (2083): 20160112. https://doi.org/10.1098/rsta.2016.0112.

Floridi, Luciano. 2017. "Infraethics—on the Conditions of Possibility of Morality". *Philosophy & Technology* 30 (4): 391–94. https://doi.org/10.1007/s13347-017-0291-1.

Floridi, Luciano. 2019. "Translating Principles into Practices of Digital Ethics: Five Risks of Being Unethical". *Philosophy & Technology* 32 (2): 185–93. https://doi.org/10.1007/s13347-019-00354-x.

Floridi, Luciano, and Josh Cowls. 2019. "A Unified Framework of Five Principles for AI in Society". *Harvard Data Science Review, June.* https://doi.org/10.1162/99608f92.8cd550d1.

Floridi, Luciano, Josh Cowls, Thomas C. King, and Mariarosaria Taddeo. 2020. "How to Design AI for Social Good: Seven Essential Factors". *Science and Engineering Ethics* 26 (3): 1771–96. https://doi.org/10.1007/s11948-020-00213-5.

Floridi, Luciano, Matthias Holweg, Mariarosaria Taddeo, Javier Amaya Silva, Jakob Mökander, and Yuni Wen. 2022. "CapAI—a Procedure for Conducting Conformity Assessment of AI Systems in Line with the EU Artificial Intelligence Act". March 23, 2022. *SSRN Electronic Journal.* https://doi.org/10.2139/ssrn.4064091.

Floridi, Luciano, and Jeff W. Sanders. 2004. "On the Morality of Artificial Agents". *Minds and Machines* 14 (3): 349–79.

Floridi, Luciano, and Mariarosaria Taddeo. 2016. "What Is Data Ethics?" *Philosophical Transactions of the Royal Society A: Mathematical, Physical and Engineering Sciences* 374 (2083): 20160360. https://doi.org/10.1098/rsta.2016.0360.

Floridi, Luciano, and Mariarosaria Taddeo. 2018. "Romans Would Have Denied Robots Legal Personhood". *Nature* 557 (7705): 309. https://doi.org/10.1038/d41586-018-05154-5.

Foreign & Commonwealth Office. 2016. "United Kingdom of Great Britain and Northern Ireland Statement to the Informal Meeting of Experts on Lethal Autonomous Weapons Systems 11–15 April 2016". https://unog.ch/80256EDD0 06B8954/(httpAssets)/44E4700A0A8CED0EC1257F940053FE3B/$file/2016_ LAWS+MX_Towardaworkingdefinition_Statements_United+Kindgom.pdf.

Foy, James. 2014. "Autonomous Weapons Systems: Taking the Human out of International Humanitarian Law". *Dalhousie Journal of Legal Studies* 23: 47–70.

FRA (European Union Agency for Fundamental Rights). 2019. "Data Quality and Artificial Intelligence—Mitigating Bias and Error to Protect Fundamental Rights". *FRA Focus*. https://fra.europa.eu/sites/default/files/fra_uploads/fra-2019-data-quality-and-ai_en.pdf.

Fraga-Lamas, Paula, Tiago M. Fernández-Caramés, Manuel Suárez-Albela, Luis Castedo, and Miguel González-López. 2016. "A Review on Internet of Things for Defense and Public Safety". *Sensors (Basel, Switzerland)* 16 (10). https://doi.org/ 10.3390/s16101644.

Freedberg, Sydeney. 2014. "NATO Hews to Strategic Ambiguity On Cyber Deterrence". 2014. https://breakingdefense.com/2014/11/natos-hews-to-strate gic-ambiguity-on-cyber-deterrence/. Accessed July 2024.

Freedman, Lawrence. 2004. *Deterrence*. Cambridge: Polity Press.

Freeman, Lindsay. 2021. "Weapons of War, Tools of Justice: Using Artificial Intelligence to Investigate International Crimes". *Journal of International Criminal Justice* 19 (1): 35–53. https://doi.org/10.1093/jicj/mqab013.

G7 Declaration. 2017. "G7 Declaration on Responsible State Behavior in Cyberspace". Lucca. http://www.mofa.go.jp/files/000246367.pdf.

Galliott, Jai. 2017. *Military Robots: Mapping the Moral Landscape*. http://www. vlebooks.com/vleweb/product/openreader?id=none&isbn=9781317096009.

Garland, David. 2003. "The Rise of Risk". In *Risk and Morality*, edited by Richard V. Ericson and Aaron Doyle. 48–86, Toronto: University of Toronto Press.

Gavaghan, Colin, Alistair Knott, James Maclaurin, John Zerilli, and Joy Liddicoat. 2019. "Government Use of Artificial Intelligence in New Zealand". Final Report on Phase 1 of the Law Foundation's Artificial Intelligence and Law in New Zealand Project. Wellington: New Zealand Law Foundation. https://www. cs.otago.ac.nz/research/ai/AI-Law/NZLF%20report.pdf.

GCHQ. 2021. "Pioneering a New National Security: The Ethics of Artificial Intelligence". *GCHQ*. https://www.gchq.gov.uk/files/GCHQAIPaper.pdf.

Geers, K. 2012. *Sun Tzu and Cyber War*. Tallinn, Estonia: Cooperative Cyber Defence Centre of Excellence.

Georgieva, Ilina, Claudio Lazo, Tjerk Timan, and Anne Fleur van Veenstra. 2022. "From AI Ethics Principles to Data Science Practice: A Reflection and a Gap Analysis Based on Recent Frameworks and Practical Experience". *AI and Ethics* 2 (4): 697–711. https://doi.org/10.1007/s43681-021-00127-3.

Glaessgen, Edward, and David Stargel. 2012. "The Digital Twin Paradigm for Future NASA and U.S. Air Force Vehicles". In *53rd AIAA/ASME/ASCE/AHS/ ASC Structures, Structural Dynamics and Materials Conference*. Honolulu, HI: American Institute of Aeronautics and Astronautics. https://doi.org/ 10.2514/6.2012-1818.

Glerup, Cecilie, and Maja Horst. 2014. "Mapping 'Social Responsibility' in Science". *Journal of Responsible Innovation* 1 (1): 31–50. https://doi.org/10.1080/23299 460.2014.882077.

Goddard, Kate, Abdul Roudsari, and Jeremy C. Wyatt. 2012. "Automation Bias: A Systematic Review of Frequency, Effect Mediators, and Mitigators". *Journal of the American Medical Informatics Association* 19 (1): 121–27.

Gomez, Steven R., Vincent Mancuso, and Diane Staheli. 2019. "Considerations for Human-Machine Teaming in Cybersecurity". In *Augmented Cognition*, edited by Dylan D. Schmorrow and Cali M. Fidopiastis, 11580: 153–68. Cham: Springer. https://doi.org/10.1007/978-3-030-22419-6_12.

Goodman, Will. 2010. "Will Goodman, Cyber Deterrence: Tougher in Theory Than in Practice?" *Strategic Studies Quarterly* 2010 (Fall): 102–35.

Grut, Chantal. 2013. "The Challenge of Autonomous Lethal Robotics to International Humanitarian Law". *Journal of Conflict and Security Law* 18 (1): 5–23.

Guastini, Roberto. 2019. "Identificazione, interpretazione dei principi costituzionali". Rome: Università degli Studi di Roma 3.

Guo, Weisi, Kristian Gleditsch, and Alan Wilson. 2018. "Retool AI to Forecast and Limit Wars". *Nature*, October 15, 2018. https://www.nature.com/articles/d41 586-018-07026-4.

Guthrie, Charles, and Michael Quinlan. 2007. *Just War: The Just War Tradition. Ethics in Modern Warfare*. London: Bloomsbury.

Habermas, Jürgen. 1990. "Discourse Ethics: Notes on a Program of Philosophical Justification". In *Moral Consciousness and Communicative Action*, translated by C. Lenhardt and S. W. Nicholsen, 43–115. Cambridge, MA: MIT Press.

Habermas, Jürgen. 1998. *The Inclusion of the Other: Studies in Political Theory.* Edited by Ciaran Cronin and Pablo De Greiff. Princeton, NJ: Princeton University Press.

Habermas, Jürgen. 2021. *The Structural Transformation of the Public Sphere: An Inquiry into a Category of Bourgeois Society*. Translated by Thomas Burger and Frederick G. Lawrence. Cambridge: Polity Press.

Haddon, Catherine. 2020. "Ministerial Accountability". Institute for Government. September 16, 2020. https://www.instituteforgovernment.org.uk/explainers/ ministerial-accountability.

Hadfield-Menell, Dylan, Smitha Milli, Pieter Abbeel, Stuart Russell, and Anca Dragan. 2020. "Inverse Reward Design". *arXiv:1711.02827 [Cs]*, October. http:// arxiv.org/abs/1711.02827.

Haggard, Simon, and Beth. A. Simmons. 1987. "Theories of International Regimes". *International Organization* 41 (03): 491.

Hala Systems. 2022. "Hala Systems". https://halasystems.com/.

Haley, Cristopher. 2013. "A Theory of Cyber Deterrence". *Georgetown Journal of International Affairs*, February. http://journal.georgetown.edu/a-theory-of-cyber-deterrence-christopher-haley/.

Harknett, Richard, J. 1996. "Information Warfare and Deterrence". *U.S. Army War College Parameters* 10 (26): 93–107.

Harknett, Richard J., and Emily O. Goldman. 2016. "The Search for Cyber Fundamentals". *Journal of Information Warfare* 15 (2): 81–88.

Harwell, Drew, and Eva Dou. 2020. "Huawei Tested AI Software That Could Recognize Uighur Minorities and Alert Police, Report Says". *Washington Post*, 2020. https://www.washingtonpost.com/technology/2020/12/08/huawei-tested-ai-software-that-could-recognize-uighur-minorities-alert-police-report-says/.

Hathaway, Oona, and Rebecca Crootof. 2012. "The Law of Cyber-attack". *California Law Review* 100 (1): 817–86.

Heath, David, Derek Allum, and Lynne Dunckley. 1994. *Introductory Logic and Formal Methods*. Henley-on-Thames: Alfred Waller.

Heath, Joseph. 2014. "Rebooting Discourse Ethics". *Philosophy & Social Criticism* 40 (9): 829–66. https://doi.org/10.1177/0191453714545340.

Heaven, Will Douglas. 2021. "DeepMind Says Its New Language Model Can Beat Others 25 Times Its Size". *MIT Technology Review*, December 8, 2021. https://www.technologyreview.com/2021/12/08/1041557/deepmind-language-model-beat-others-25-times-size-gpt-3-megatron/.

Hepenstal, Sam, Leishi Zhang, Neesha Kodagoda, and B. L. William Wong. 2020. "Pan: Conversational Agent for Criminal Investigations". In *Proceedings of the 25th International Conference on Intelligent User Interfaces Companion*, edited by Fabio Partenò, 134–35. New York, NY: Association for Computing Machinery.

Hersh, Seymour. 2000. "Overwhelming Force: What Happened in the Final Days of the Gulf War?" *New Yorker*, May 22, 2000.

Heyns, Christof. 2014. "Autonomous Weapons Systems and Human Rights Law". Presentation made at the Informal Expert Meeting Organized by the State Parties to the Convention on Certain Conventional Weapons, May 13–16, 2014, Geneva.

Heyns, Christof. 2016a. "Autonomous Weapons Systems: Living a Dignified Life and Dying a Dignified Death". *In Autonomous Weapons Systems: Law, Ethics, Policy*, edited by Nehal Bhuta, Susanne Beck, Robin Geiß, Hin-Yan Liu, and Claus Kreß, 3–19. Cambridge: Cambridge University Press.

Heyns, Christof. 2016b. "Human Rights and the Use of Autonomous Weapons Systems (AWS) during Domestic Law Enforcement". *Human Rights Quarterly* 38 (2): 350–78. https://doi.org/10.1353/hrq.2016.0034.

High-Level Expert Group on Artificial Intelligence. 2019. *Ethics Guidelines for Trustworthy AI*. Brussels: European Commission. https://www.aepd.es/sites/default/files/2019-12/ai-ethics-guidelines.pdf.

Hin-Yan Liu. 2016. "Refining Responsibility: Differentiating Two Types of Responsibility Issues Raised by Autonomous Weapons Systems". *In Autonomous Weapons Systems: Law, Ethics, Policy*, edited by Nehal Bhuta, Susanne Beck, Robin Geiß, Hin-Yan Liu, and Claus Kreß, 325–44. Cambridge: Cambridge University Press.

HM Government. 2022. "National Cyber Strategy". 130. https://www.gov.uk/government/publications/national-cyber-strategy-2022. Accessed July 2024.

Hoare, C. A. R. 1972. "Notes on Data Structuring". In *Structured Programming*, edited by O. J. Dahl, E. W. Dijkstra, and C. A. R. Hoare, 83–174. London: Academic Press. http://dl.acm.org/citation.cfm?id=1243380.1243382.

Hoffman, Wyatt. 2021. "AI and the Future of Cyber Competition". Center for Security and Emerging Technology. https://doi.org/10.51593/2020CA007.

Holland Michel, Arthur. 2020a. "The Black Box, Unlocked | UNIDIR". 2020. https://unidir.org/publication/black-box-unlocked.

Holland Michel, Arthur. 2020b. "The Black Box, Unlocked: Predictability and Understandability in Military AI". United Nations Institute for Disarmament Research. https://doi.org/10.37559/SecTec/20/AI1.

Hollis, Duncan B. 2011. "An E-SOS for Cyberspace". *Harvard International Law Journal* 52 (373): 374–75.

House of Lords. 2019. "Autonomous Weapons: Questions for Ministry of Defence UIN HL15333". April 24, 2019. https://questions-statements.parliament.uk/written-questions/detail/2019-04-24/HL15333.

Hüllermeier, Eyke, and Willem Waegeman. 2021. "Aleatoric and Epistemic Uncertainty in Machine Learning: An Introduction to Concepts and Methods". *Machine Learning* 110 (3): 457–506. https://doi.org/10.1007/s10994-021-05946-3.

Hume, David. 2009. *A Treatise of Human Nature*. Edited by David Fate Norton. Oxford: Oxford University Press.

Hurka, Thomas. 2005. "Proportionality in the Morality of War". *Philosophy & Public Affairs* 33 (1): 34–66.

Hurka, Thomas. 2008. "Proportionality and Necessity". In *War: Essays in Political Philosophy*, edited by Larry May and Emily Crookston, 127–44. Cambridge: Cambridge University Press.

Ian Clark. 2015. *Waging War: A New Philosophical Introduction*. Oxford: Oxford University Press.

Iasiello, Emilio. 2014. "Is Cyber Deterrence an Illusory Course of Action?" *Journal of Strategic Security* 7 (1): 54–67.

IBM. 2021. "What Is Data Labeling?" August 12, 2021. https://www.ibm.com/cloud/learn/data-labeling.

Independent Surveillance Review. 2015. "A Democratic Licence to Operate: Report of the Independent Surveillance Review". London: Royal United Services Institute for Defence Studies. https://static.rusi.org/20150714_whr_2-15_a_democratic_licence_to_operate.pdf.

Insikt Group. 2022. "HermeticWiper and PartyTicket Targeting Computers in Ukraine". March 2, 2022. https://go.recordedfuture.com/hubfs/reports/mtp-2022-0302.pdf.

International Committee of the Red Cross. 2016. "Views of the ICRC on Autonomous Weapon Systems". April. https://www.icrc.org/en/document/views-icrc-autonomous-weapon-system.

International Committee of the Red Cross. 2018. "Ethics and Autonomous Weapon Systems: An Ethical Basis for Human Control?", Geneve, Switzerland.

International Committee of the Red Cross. 2019. "Autonomy, Artificial Intelligence and Robotics: Technical Aspects of Human Control". https://www.icrc.org/en/document/autonomy-artificial-intelligence-and-robotics-technical-aspects-human-control.

International Committee of the Red Cross. 2020. "Treaties, States Parties, and Commentaries - St Petersburg Declaration Relating to Explosive Projectiles, 1868 - Declaration." Geneva: International Committee of the Red Cross. https://ihl-databases.icrc.org/applic/ihl/ihl.nsf/Article.xsp?action=openD

ocument&documentId=568842C2B90F4A29C12563CD0051547C.   Accessed December 2, 2022.

International Committee of the Red Cross. 2021. "ICRC Position on Autonomous Weapon Systems & Background Paper". Geneva: International Committee of the Red Cross.

International Military Tribunal (Nuremberg). 1947. "Judgment and Sentences, October 1, 1946". *American Journal of International Law* 41: 172–306.

International Security Advisory Board. 2014. "A Framework for International Cyber Stability". US Department of State. http://goo.gl/azdM0B.

Ish, Daniel, Jared Ettinger, and Christopher Ferris. 2021. "Evaluating the Effectiveness of Artificial Intelligence Systems in Intelligence Analysis". Rand Corporation. https://www.rand.org/pubs/research_reports/RRA464-1.html.

Jacky, Jonathan. 1997. *The Way of Z: Practical Programming with Formal Methods.* Cambridge: Cambridge University Press.

Jagielski, Matthew, Alina Oprea, Battista Biggio, Chang Liu, Cristina Nita-Rotaru, and Bo Li. 2018. "Manipulating Machine Learning: Poisoning Attacks and Countermeasures for Regression Learning". *arXiv:1804.00308 [Cs]*, April. http://arxiv.org/abs/1804.00308.

Jain, Neha. 2016. "Autonomous Weapons Systems: New Frameworks for Individual Responsibility". In *Autonomous Weapons Systems: Law, Ethics, Policy*, edited by Nehal Bhuta, Susanne Beck, Robin Geiß, Hin-Yan Liu, and Claus Kreß, 303–24. Cambridge: Cambridge University Press.

Japanese Society for Artificial Intelligence. 2017. "Ethical Guidelines". http://ai-elsi. org/wp-content/uploads/2017/05/JSAI-Ethical-Guidelines-1.pdf.

Jensen, Eric Talbot. 2009. "Cyber Warfare and Precautions against the Effects of Attacks". *Texas Law Review* 88 (1533): 1534–69.

Jensen, Eric Talbot. 2012. "Cyber Deterrence". *Emory International Law Review* 26 (2): 773–824.

Jervis, Robert. 1979. "Deterrence Theory Revisited". *World Politics* 31 (2): 289–324. https://doi.org/10.2307/2009945.

Jervis, Robert. 1988. "Realism, Game Theory, and Cooperation". *World Politics* 40 (3): 317–49. https://doi.org/10.2307/2010216.

Jia, Yifan, Christopher M. Poskitt, Jun Sun, and Sudipta Chattopadhyay. 2022. "Physical Adversarial Attack On a Robotic Arm". *IEEE Robotics and Automation Letters* 7 (4): 9334–41.

Johnson, Aaron M., and Sidney Axinn. 2013. "The Morality of Autonomous Robots". *Journal of Military Ethics* 12 (2): 129–41. https://doi.org/10.1080/15027570.2013.818399.

Johnston, Rob. 2005. *Analytic Culture in the US Intelligence Community: An Ethnographic Study.* Washington, DC: Center for the Study of Intelligence, Central Intelligence Agency. https://web.archive.org/web/20070613143919/https://www.cia.gov/library/center-for-the-study-of-intelligence/csi-publicati ons/books-and-monographs/analytic-culture-in-the-u-s-intelligence-commun ity/chapter_1.htm.

Justice and Home Affairs Committee. 2022. "Technology Rules? The Advent of New Technologies in the Justice System". HLPaper180. Westminster: House of Lords.

Kamm, F. M. 2004. "Failures of Just War Theory: Terror, Harm, and Justice". *Ethics* 114 (4): 650–92. https://doi.org/10.1086/383441.

Kania, Elsa B. 2018a. "China's Embrace of AI: Enthusiasm and Challenges – European Council on Foreign Relations". *ECFR* (blog), November 6, 2018. https://ecfr.eu/article/commentary_chinas_embrace_of_ai_enthusiasm_and_challenges/.

Kania, Elsa B. 2018b. "China's Strategic Ambiguity and Shifting Approach to Lethal Autonomous Weapons Systems". *Lawfare* (blog), April 17, 2018. https://www.lawfareblog.com/chinas-strategic-ambiguity-and-shifting-approach-lethal-autonomous-weapons-systems.

Kant, Immanuel. 2019. *Grundlegung zur Metaphysik der Sitten (Großdruck)*. Edited by Theodor Borken. Berlin: Henricus.

Kasher, Asa. 2007. "The Principle of Distinction". *Journal of Military Ethics* 6 (2): 152–67.

Kastenberg, Joshua E. 2009. "Changing the Paradigm of Internet Access from Government Information Systems: A Solution to the Need for the DoD to Take Time-Sensitive Action on the Niprnet". *Air Force Law Review* 64: 175: 175–210.

Kaurin, Pauline. 2010. "With Fear and Trembling: An Ethical Framework for Non-lethal Weapons". *Journal of Military Ethics* 9 (1): 100–114. https://doi.org/10.1080/15027570903523057.

Kaurin, Pauline. 2015. "And Next Please: The Future of the NLW Debate International Regulation of Emerging Military Technologies". *Case Western Reserve Journal of International Law* 47 (1): 217–28.

Kelion, Leo. 2021. "Huawei Patent Mentions Use of Uighur-Spotting Tech". *BBC News*, January 13, 2021, sec. Technology. https://www.bbc.com/news/technology-55634388.

Kelly, Erin I. 2012. "What Is an Excuse?" In *Blame*, edited by D. Justin Coates and Neal A. Tognazzini, 244–62. New York: Oxford University Press. https://doi.org/10.1093/acprof:oso/9780199860821.003.0013.

Kelly, Jonathan, Michael DeLaus, Erik Hemberg, and Una-May O'Reilly. 2019. "Adversarially Adapting Deceptive Views and Reconnaissance Scans on a Software Defined Network". In *FIP/IEEE Symposium on Integrated Network and Service Management (IM)*, edited by Nur Zincir-Heywood, Idilio Drago, and Robert Harper, 49–54. Piscataway, NJ: IEEE.

Khosrow-Pour D.B.A., Mehdi, ed. 2021. *Encyclopedia of Information Science and Technology*. 5th ed. Hersey, PA: IGI Global. https://doi.org/10.4018/978-1-7998-3479-3.

Khoury, Andrew C. 2018. "The Objects of Moral Responsibility". *Philosophical Studies* 175 (6): 1357–81. https://doi.org/10.1007/s11098-017-0914-5.

Kim, Scott Y. H., Ian F. Wall, Aimee Stanczyk, and Raymond De Vries. 2009. "Assessing the Public's Views in Research Ethics Controversies: Deliberative Democracy and Bioethics as Natural Allies". *Journal of Empirical Research on Human Research Ethics* 4 (4): 3–16. https://doi.org/10.1525/jer.2009.4.4.3.

King, Tariq M., Jason Arbon, Dionny Santiago, David Adamo, Wendy Chin, and Ram Shanmugam. 2019. "AI for Testing Today and Tomorrow: Industry Perspectives". In *2019 IEEE International Conference on Artificial Intelligence*

Testing (AITest), 81–88. Newark, CA: IEEE. https://doi.org/10.1109/AIT est.2019.000-3.

Kirat, Dhilung, Jiyong Jang, and Marc Ph. Stoecklin. 2018. "DeepLocker: Concealing Targeted Attacks with AI Locksmithing". IBM. https://i.blackhat.com/us-18/ Thu-August-9/us-18-Kirat-DeepLocker-Concealing-Targeted-Attacks-with-AI-Locksmithing.pdf.

Klamm, J., C. Dominguez, B. Yost, P. McDermott, and M. Lenox. 2019. "Partnering with Technology: The Importance of Human Machine Teaming in Future MDC2 Systems". In *Artificial Intelligence and Machine Learning for Multi-Domain Operations Applications*, 11006:259–66. SPIE. https://doi.org/10.1117/ 12.2518750.

Kniep, Ronja. 2019. "Another Layer of Opacity: How Spies Use AI and Why We Should Talk about It". *About:Intel* (blog), December 20, 2019. https://aboutin tel.eu/how-spies-use-ai/.

Knight, Will. 2022. *Wired*, March 17, 2022. https://www.wired.com/story/ai-dro nes-russia-ukraine/.

Konaev, Margarita, and Husanjot Chahal. 2021. "Building Trust in Human-Machine Teams". https://www.brookings.edu/techstream/building-trust-in-human-machine-teams/.

Korpela, Christopher. 2017. "Report of the 2017 Group of Governmental Experts on Lethal Autonomous Weapons Systems (LAWS)". *CCW/GGE.1/2017/CRP.1*. Geneva: United Nations Office for Disarmament Affairs. https://docs-library. unoda.org/Convention_on_Certain_Conventional_Weapons_-Group_of_ Governmental_Experts_on_Lethal_Autonomous_Weapons_Systems_(2023)/ CCW_GGE1_2023_CRP.1_0.pdf.

Kott, Alexander. 2018. "Intelligent Autonomous Agents Are Key to Cyber Defense of the Future Army Networks". *arXiv:1812.08014 [Cs]*, December. http://arxiv. org/abs/1812.08014.

Kott, Alexander, Paul Théron, Luigi V. Mancini, Edlira Dushku, Agostino Panico, Martin Drašar, Benoît LeBlanc, et al. 2020. "An Introductory Preview of Autonomous Intelligent Cyber-defense Agent Reference Architecture, Release 2.0". *Journal of Defense Modeling and Simulation: Applications, Methodology, Technology* 17 (1): 51–54. https://doi.org/10.1177/1548512919886163.

Krishnan, Armin. 2009. *Killer Robots: Legality and Ethicality of Autonomous Weapons*. Burlington, VT: Ashgate.

Krishnan, Maya. 2020. "Against Interpretability: A Critical Examination of the Interpretability Problem in Machine Learning". *Philosophy & Technology* 33 (3): 487–502. https://doi.org/10.1007/s13347-019-00372-9.

Kugler, Richard. 2009. "Deterrence of Cyber Attacks". In *Cyberpower and National Security*, edited by Franklin Kramer, Stuart Starr, and Larry Wentz, 309–42. Washington, DC: National Defense University.

La Fors, Karolina, Bart Custers, and Esther Keymolen. 2019. "Reassessing Values for Emerging Big Data Technologies: Integrating Design-Based and Application-Based Approaches". *Ethics and Information Technology* 21 (3): 209–26. https://doi.org/10.1007/s10676-019-09503-4.

Laird, John, Charan Ranganath, and Samuel Gershman. 2019. "Future Directions in Human Machine Teaming Workshop". https://basicresearch.defense.gov/

Portals/61/Future%20Directions%20in%20Human%20Machine%20Team
ing%20Workshop%20report%20%20%28for%20public%20release%29.pdf.

Lan, Tang, Zhang Xin, Harry Raduege Jr., Dmitry Grigoriev, Pavan Duggal, and
Stein Schjølberg. 2010. *Global Cyber Deterrence Views from China, the U.S.,
Russia, India, and Norway*. New York: EastWest Institute.

Lango, John W. 2010. "Nonlethal Weapons, Noncombatant Immunity, and
Combatant Nonimmunity: A Study of Just War Theory". *Philosophia* 38
(3): 475–97. https://doi.org/10.1007/s11406-009-9231-3.

Lavin, Alexander, Hector Zenil, Brooks Paige, David Krakauer, Justin Gottschlich,
Tim Mattson, Anima Anandkumar, et al. 2021. "Simulation Intelligence: Towards
a New Generation of Scientific Methods". *arXiv:2112.03235 [Cs], December*.
http://arxiv.org/abs/2112.03235.

Lebreton, Gilles. 2021. "Report of the Committee on Legal Affairs to the European
Parliament". https://www.europarl.europa.eu/doceo/document/A-9-2021-
0001_EN.html#

Levy, Neil. 2008. "The Responsibility of the Psychopath Revisited". *Philosophy,
Psychiatry, & Psychology* 14 (2): 129–38. https://doi.org/10.1353/ppp.0.0003.

Liao, Cong, Haoti Zhong, Anna Squicciarini, Sencun Zhu, and David Miller.
2018. "Backdoor Embedding in Convolutional Neural Network Models via
Invisible Perturbation". *arXiv:1808.10307 [Cs, Stat], August*. http://arxiv.org/
abs/1808.10307.

Libicki, Martin C. 1997. "Defending Cyberspace and Other Metaphors".

Libicki, Martin C. 2009. *Cyberdeterrence and Cyberwar*. Santa Monica, CA: Rand.
http://www.rand.org/pubs/monographs/MG877.html.

Lieblich, Eliav, and Eyal Benvenisti. 2016. "The Obligation to Exercise Discretion
in Warfare: Why Autonomous Weapons Systems Are Unlawful". In *Autonomous
Weapons Systems: Law, Ethics, Policy*, edited by Nehal Bhuta, Susanne Beck,
Robin Geiß, Hin-Yan Liu, and Claus Kreß, 245–83. Cambridge: Cambridge
University Press.

Lin, Herbert. 2012. "Cyber Conflict and International Humanitarian Law".
International Review of the Red Cross 94 (886): 515–31. https://doi.org/10.1017/
S1816383112000811.

Lippert-Rasmussen, Kasper. 2014. "Just War Theory, Intentions, and the
Deliberative Perspective Objection". In *How We Fight: Ethics in War*, edited by
Helen Frowe and Gerald Lang, 138–54. Oxford: Oxford University Press.

List, Christian, and Philip Pettit. 2011. *Group Agency*. New York: Oxford University
Press. https://doi.org/10.1093/acprof:oso/9780199591565.001.0001.

Llorens, Albertina Albors. 1999. "The European Court of Justice, More Than a
Teleological Court". *Cambridge Yearbook of European Legal Studies* 2: 373–98.
https://doi.org/10.5235/152888712802815789.

Lo, Chris. 2015. "Safer with Data: Protecting Pakistan's Schools with Predictive
Analytics". *Army Technology*, November 8, 2015. https://www.army-technology.
com/features/featuresafer-with-data-protecting-pakistans-schools-with-pre
dictive-analytics-4713601/.

Lopez, Todd. 2022. "Simplified Human/Machine Interfaces Top List of Critical
DOD Technologies". https://www.defense.gov/News/News-Stories/Article/

Article/2904627/simplified-humanmachine-interfaces-top-list-of-critical-dod-technologies/.

"Losing Humanity: The Case against Killer Robots". 2012. Human Rights Watch. November 19, 2012. https://www.hrw.org/report/2012/11/19/losing-human ity/case-against-killer-robots.

Lysaght, Robert J., Regina Harris, and William Kelly. 1988. "Artificial Intelligence for Command and Control". Willow Grove, PA: Analytics. https://apps.dtic.mil/docs/citations/ADA229342.

Makarius, Erin E., Debmalya Mukherjee, Joseph D. Fox, and Alexa K. Fox. 2020. "Rising with the Machines: A Sociotechnical Framework for Bringing Artificial Intelligence into the Organization". *Journal of Business Research* 120 (November): 262–73. https://doi.org/10.1016/j.jbusres.2020.07.045.

Mäntymäki, Matti, Matti Minkkinen, Teemu Birkstedt, and Mika Viljanen. 2022. "Defining Organizational AI Governance". *AI and Ethics* 2 (4): 603–9. https://doi.org/10.1007/s43681-022-00143-x.

Marchant, Gary E., Braden Allenby, Ronald Arkin, and Edward T. Barrett. 2011. "International Governance of Autonomous Military Robots". *Columbia Science and Technology Law Review* 12: 272–316.

Marcum, Richard A., Curt H. Davis, Grant J. Scott, and Tyler W. Nivin. 2017. "Rapid Broad Area Search and Detection of Chinese Surface-to-Air Missile Sites Using Deep Convolutional Neural Networks". *Journal of Applied Remote Sensing* 11 (4): 042614. https://doi.org/10.1117/1.JRS.11.042614.

Matsumoto, Masakazu. 2020. "Amoral Realism or Just War Morality? Disentangling Different Conceptions of Necessity". *European Journal of International Relations* 26 (4): 1084–105. https://doi.org/10.1177/1354066120910233.

Matthias, Andreas. 2004. "The Responsibility Gap: Ascribing Responsibility for the Actions of Learning Automata". *Ethics and Information Technology* 6 (3): 175–83. https://doi.org/10.1007/s10676-004-3422-1.

McCarthy, Thomas. 1995. "Practical Discourse: On the Relation of Morality to Politics". *Revue Internationale de Philosophie* 49 (194): 461–81.

McConnell, Mike. 2010. "Mike McConnell on How to Win the Cyber-War We're Losing". February 28, 2010. http://www.washingtonpost.com/wp-dyn/content/article/2010/02/25/AR2010022502493.html.

McIntyre, Alison. 2004. "Doctrine of Double Effect". July. https://stanford.library.sydney.edu.au/entries/double-effect/.

McKendrick, Kathleen. 2019. *Artificial Intelligence Prediction and Counterterrorism*. London: Chatham House. https://www.chathamhouse.org/sites/default/files/2019-08-07-AICounterterrorism.pdf.

McMahan, Jeff. 2006. "On the Moral Equality of Combatants". *Journal of Political Philosophy* 14 (4): 377–93.

McMahan, Jeff. 2009. *Killing in War*. Oxford: Oxford University Press.

McMahan, Jeff. 2010. "The Just Distribution of Harm between Combatants and Noncombatants". *Philosophy & Public Affairs* 38 (4): 342–79. https://doi.org/10.1111/j.1088-4963.2010.01196.x.

McMahan, Jeff. 2011. "Who Is Morally Liable to Be Killed in War?" *Analysis* 71 (3): 544–59.

McMahan, Jeff, and Robert McKim. 1993. "The Just War and the Gulf War". *Canadian Journal of Philosophy* 23 (4): 501–41.

McNeese, Nathan J., Beau G. Schelble, Lorenzo Barberis Canonico, and Mustafa Demir. 2021. "Who/What Is My Teammate? Team Composition Considerations in Human-AI Teaming". *arXiv:2105.11000 [Cs], May*. http://arxiv.org/abs/2105.11000.

Meisels, Tamar. 2018. *Contemporary Just War: Theory and Practice*. London: Routledge.

Miller, Seumas. 2018. *Dual Use Science and Technology, Ethics and Weapons of Mass Destruction*. New York: Springer.

Ministry of Defence. 2011. "Joint Service Manual of The Law of Armed Conflict (JSP 383)". https://assets.publishing.service.gov.uk/government/uploads/system/uploads/attachment_data/file/27874/JSP3832004Edition.pdf.

Ministry of Defence. 2018a. "Unmanned Aircraft Systems (JDP 0-30.2)". https://www.gov.uk/government/publications/unmanned-aircraft-systems-jdp-0-302.

Ministry of Defence. 2018b. "Human-Machine Teaming (JCN 1/18)". https://www.gov.uk/government/publications/human-machine-teaming-jcn-118.

Ministry of Defence. 2022. "Ambitious, Safe, Responsible: Our Approach to the Delivery of AI-Enabled Capability in Defence". https://assets.publishing.service.gov.uk/government/uploads/system/uploads/attachment_data/file/1082991/20220614-Ambitious_Safe_and_Responsible.pdf.

Mirsky, Yisroel, Tom Mahler, Ilan Shelef, and Yuval Elovici. 2019. "CT-GAN: Malicious Tampering of 3D Medical Imagery Using Deep Learning". *ResearchGate*. https://www.researchgate.net/publication/330357848_CT-GAN_Malicious_Tampering_of_3D_Medical_Imagery_using_Deep_Learning/figures?lo=1.

Mitchell, Kwasi, Joe Mariani, Adam Routh, Akash Keyal, and Alex Mirkow. 2019. *The Future of Intelligence Analysis: A Task-Level View of the Impact of Artificial Intelligence on Intel Analysis*. Washington, DC: Deloitte.

Mitchell, Margaret, Simone Wu, Andrew Zaldivar, Parker Barnes, Lucy Vasserman, Ben Hutchinson, Elena Spitzer, Inioluwa Deborah Raji, and Timnit Gebru. 2019. "Model Cards for Model Reporting". In *Proceedings of the Conference on Fairness, Accountability, and Transparency—FAT* '19*, 220–29. https://doi.org/10.1145/3287560.3287596.

Mökander, Jakob, and Luciano Floridi. 2021. "Ethics-Based Auditing to Develop Trustworthy AI". *Minds and Machines*, February. https://doi.org/10.1007/s11023-021-09557-8.

Moor, James H. 1985. "What Is Computer Ethics?*". *Metaphilosophy* 16 (4): 266–75. https://doi.org/10.1111/j.1467-9973.1985.tb00173.x.

Moore, Cristopher. 1990. "Unpredictability and Undecidability in Dynamical Systems". *Physical Review Letters* 64 (20): 2354–57. https://doi.org/10.1103/PhysRevLett.64.2354.

Morgan, Patrick M. 2003. *Deterrence Now*. Cambridge: Cambridge University Press.

Morgan, Patrick M. 2010. "Applicability of Traditional Deterrence Concepts and Theory to the Cyber Realm". In *Proceedings of a Workshop on Deterring*

Cyberattacks: Informing Strategies and Developing Options for U.S. Policy, 55–76. Washington, DC: National Academic Press.

Morley, Jessica, Josh Cowls, Mariarosaria Taddeo, and Luciano Floridi. 2020. "Ethical Guidelines for COVID-19 Tracing Apps". *Nature* 582: 29–31.

Morley, Jessica, Anat Elhalal, Francesca Garcia, Libby Kinsey, Jakob Mökander, and Luciano Floridi. 2021. "Ethics as a Service: A Pragmatic Operationalisation of AI Ethics". *Minds and Machines* 31 (2): 239–56. https://doi.org/10.1007/s11 023-021-09563-w.

Morley, Jessica, Luciano Floridi, Libby Kinsey, and Anat Elhalal. 2020. "From What to How: An Initial Review of Publicly Available AI Ethics Tools, Methods and Research to Translate Principles into Practices". *Science and Engineering Ethics* 26 (4): 2141–68. https://doi.org/10.1007/s11948-019-00165-5.

Moseley, Alexander. 2011. "Just War Theory". In *The Encyclopedia of Peace Psychology*, edited by D. J. Christie. John Wiley & Sons. https://doi.org/10.1002/ 9780470672532.wbepp144.

Mueller, John. 1995. "The Perfect Enemy: Assessing the Gulf War". *Security Studies* 5 (1): 77–117. https://doi.org/10.1080/09636419508429253.

Musiolik, Thomas Heinrich, and Adrian David Cheok, eds. 2021. *Analyzing Future Applications of AI, Sensors, and Robotics in Society: Advances in Computational Intelligence and Robotics*. Hersey, PA: IGI Global. https://doi.org/10.4018/ 978-1-7998-3499-1.

Nagel, Thomas. 1972. "War and Massacre". *Philosophy and Public Affairs* 1 (Winter): 123–44.

National Academies of Sciences, Engineering, and Medicine, Committee on Human-System Integration Research Topics for the 711th Human, Performance Wing of the Air Force Research Laboratory, Board on Human-Systems Integration, Division of Behavioral and Social Sciences and Education, and Board on Human-Systems Integration. 2022. *Human-AI Teaming: State-of-the-Art and Research Needs*. Washington, DC: National Academies Press. https:// doi.org/10.17226/26355.

National Security Agency. 2012. "U (SIGINT Strategy". February 23, 2012. In "A Strategy for Surveillance Powers". *New York Times*, November 23, 2013. http:// www.nytimes.com/interactive/2013/11/23/us/politics/23nsa-sigint-strategy- document.html.

NATO. 2020. "AAP-06 Edition 2020: NATO Glossary of Terms and Definitions". NATO Standardization Office.

NATO Cooperative Cyber Defence Centre of Excellence. 2013. *Tallinn Manual on the International Law Applicable to Cyber Warfare: Prepared by the International Group of Experts at the Invitation of the NATO Cooperative Cyber Defence Centre of Excellence*. Cambridge: Cambridge University Press.

Nelkin, Dana Kay. 2011. *Making Sense of Freedom and Responsibility*. New York: Oxford University Press.

Nguyen, Anh M., J. Yosinski, and J. Clune. 2015. "Deep Neural Networks Are Easily Fooled: High Confidence Predictions for Unrecognizable Images". In *2015 IEEE Conference on Computer Vision and Pattern Recognition (CVPR)*. Boston, MA. https://doi.org/10.1109/CVPR.2015.7298640.

NIST. 2022. "AI Risk Management Framework: Initial Draft". March 17, 2022. https://www.nist.gov/system/files/documents/2022/03/17/AI-RMF-1stdr aft.pdf.

Niu, Yaru, Rohan Paleja, and Matthew Gombolay. 2021. "Multi-agent Graph-Attention Communication and Teaming". 10.

Noor, Umara, Zahid Anwar, Tehmina Amjad, and Kim-Kwang Raymond Choo. 2019. "A Machine Learning-Based FinTech Cyber Threat Attribution Framework Using High-Level Indicators of Compromise". *Future Generation Computer Systems* 96 (July): 227–42. https://doi.org/10.1016/j.future.2019.02.013.

Norway. 2017. "CCW Group of Governmental Experts on Lethal Autonomous Weapons Systems: General Statement by Norway". https://docs-library.unoda.org/Convention_on_Certain_Conventional_Weapons_-_Group_of_Go vernmental_Experts_(2017)/2017_GGE%2BLAWS_Statement_Norway. pdf. Access July 2024. https://docs-library.unoda.org/Convention_on_Certai n_Conventional_Weapons_-_Group_of_Governmental_Experts_(2017)/ 2017_GGE%2BLAWS_Statement_Norway.pdf

Nurick, Lester. 1945. "The Distinction between Combatant and Noncombatant in the Law of War". *American Journal of International Law* 39 (4): 680–97. https://doi.org/10.2307/2193409.

Nye, Joseph S. 2011. "Nuclear Lessons for Cyber Security?" *Strategic Studies Quarterly* 5 (4): 11–38.

O'Connell, Mary Ellen. 2012. "Cyber Security without Cyber War". *Journal of Conflict and Security Law* 17 (2): 187–209. https://doi.org/10.1093/jcsl/krs017.

O'Connell, Mary Ellen. 2014. "The American Way of Bombing: How Legal and Ethical Norms Change". In *The American Way of Bombing: Changing Ethical and Legal Norms, from Flying Fortresses to Drones*, edited by Matthew Evangelista and Henry Shue, 1–24. Ithaca, NY: Cornell University Press.

Office of the Secretary of Defense. 2017. "Department of Defense Fiscal Year (FY) 2017 Request for Additional Appropriations".

Ohlin, Jens David, and Larry May. 2016. *Necessity in International Law.* Oxford: Oxford University Press.

Omand, David, and Mark Phythian. 2018. *Principled Spying: The Ethics of Secret Intelligence.* Oxford: Oxford University Press.

O'Neill, Thomas, Nathan McNeese, Amy Barron, and Beau Schelble. 2020. "Human-Autonomy Teaming: A Review and Analysis of the Empirical Literature". *Human Factors*, October, 001872082096086. https://doi.org/ 10.1177/0018720820960865.

OpenAI. 2019. "Better Language Models and Their Implications". OpenAI (blog), February 14, 2019. https://openai.com/blog/better-language-models/.

Orend, Brian. 2001. "Just and Lawful Conduct in War: Reflections on Michael Walzer". *Law and Philosophy* 20 (1): 1–30. https://doi.org/10.2307/3505049.

Orend, Brian. 2019. *War and Political Theory.* Cambridge: Polity.

Owens, William A., Kenneth W. Dam, and Herbert Lin, eds. 2009. *Technology, Policy, Law, and Ethics Regarding U.S. Acquisition and Use of Cyberattack Capabilities.* Washington, DC: National Academies Press.

Payne, Kenneth. 2021. *I, Warbot: The Dawn of Artificially Intelligent Conflict.* London: Hurst & Company.

Pellerin, Cheryl. 2017. "Project Maven Industry Day Pursues Artificial Intelligence for DoD Challenges". US Department of Defense. https://www.defense.gov/News/News-Stories/Article/Article/1356172/project-maven-industry-day-pursues-artificial-intelligence-for-dod-challenges/.

Perry, Stephen R. 1995. "Risk, Harm, and Responsibility". In *Philosophical Foundations of Tort Law*, edited by David G. Owen, 321–46. Oxford: Oxford University Press. https://watermark.silverchair.com/acprof-9780198265795-chapter-15.pdf?t.

Peters, Dorian. 2019. "Beyond Principles: A Process for Responsible Tech". *The Ethics of Digital Experience* (blog), May 14, 2019. https://medium.com/ethics-of-digital-experience/beyond-principles-a-process-for-responsible-tech-aefc9 21f7317.

Possony, Stefan T. 1946. "Atomic Power and World Order". *Review of Politics* 8 (4): 533–35.

Powell, Robert. 2008. *Nuclear Deterrence Theory: The Search for Credibility*. Cambridge: Cambridge University Press.

Raaijmakers, Stephan. 2019. "Artificial Intelligence for Law Enforcement: Challenges and Opportunities". *IEEE Security & Privacy* 17 (5): 74–77.

Rae, Jack, Geoffrey Irving, and Laura Weidinger. 2021. "Language Modelling at Scale: Gopher, Ethical Considerations, and Retrieval". DeepMind *(blog)*, December 8, 2021. https://deepmind.com/blog/article/language-modell ing-at-scale.

Ramsey, Paul. 2002. *The Just War: Force and Political Responsibility*. Lanham, MD: Rowman & Littlefield.

Rassler, Don. 2021. "Data, AI, and the Future of U.S. Counterterrorism: Building an Action Plan". *CTC Sentinel* 14 (8): 31–44.

Rattray, Gregory J. 2009. "An Environmental Approach to Understanding Cyberpower," in Kramer, Cited, 253-274, Esp. 256". In *Cyberpower and National Security*, edited by Stuart S. Kramer and Lerry K. Wentz, 253–74. Washington, DC: National Defense University Press.

Rawls, John. 2005. *A Theory of Justice*. Cambridge, MA: Belknap Press.

République Française. 2016. "Working Paper of France: 'Characterization of A Laws'". In *Meeting of Experts on Lethal Autonomous Weapons Systems (LAWS)*. https://unog.ch/80256EDD006B8954/(httpAssets)/5FD844883B46FEACC 1257F8F00401FF6/$file/2016_LAWSMX_CountryPaper_France+Characte rizationofaLAWS.pdf.

Rice, H. G. 1956. "On Completely Recursively Enumerable Classes and Their Key Arrays". *Journal of Symbolic Logic* 21 (3): 304–8. https://doi.org/10.2307/2269105.

Rigaki, Maria, and Ahmed Elragal. 2017. "Adversarial Deep Learning against Intrusion Detection Classifiers". Master's thesis, Luleå University of Technology.

Robbins, Martin. 2016. "Has a Rampaging AI Algorithm Really Killed Thousands in Pakistan?" *The Guardian*, February 18, 2016, sec. Science. https://www.theg uardian.com/science/the-lay-scientist/2016/feb/18/has-a-rampaging-ai-algori thm-really-killed-thousands-in-pakistan.

Roberts, Huw, Josh Cowls, Jessica Morley, Mariarosaria Taddeo, Vincent Wang, and Luciano Floridi. 2020. "The Chinese Approach to Artificial Intelligence: An

Analysis of Policy, Ethics, and Regulation". *AI & Society*, June. https://doi.org/10.1007/s00146-020-00992-2.

Robinette, Paul, Ayanna M. Howard, and Alan R. Wagner. 2017. "Effect of Robot Performance on Human–Robot Trust in Time-Critical Situations". *IEEE Transactions on Human-Machine Systems* 47 (4): 425–36. https://doi.org/10.1109/THMS.2017.2648849.

Robinette, Paul, Wenchen Li, Robert Allen, Ayanna M. Howard, and Alan R. Wagner. 2016. "Overtrust of Robots in Emergency Evacuation Scenarios". In *2016 11th ACM/IEEE International Conference on Human-Robot Interaction (HRI)*,101–108. Christchurch, New Zealand: IEEE.

Roff, Heather M. 2014. "The Strategic Robot Problem: Lethal Autonomous Weapons in War". *Journal of Military Ethics* 13 (3): 211–27. https://doi.org/10.1080/15027570.2014.975010.

Roff, Heather M. 2015. "Lethal Autonomous Weapons and *Jus ad Bellum* Proportionality". *Case Western Reserve Journal of International Law* 47 (1): 37–52.

Roff, Heather M. 2020a. *Uncomfortable Ground Truths: Predictive Analytics and National Security*. Washington, DC: Brookings Institute.

Roff, Heather M. 2020b. "Forecasting and Predictive Analytics: A Critical Look at the Basic Building Blocks of a Predictive Model". *Brookings* (blog), September 11, 2020. https://www.brookings.edu/techstream/forecasting-and-predictive-analytics-a-critical-look-at-the-basic-building-blocks-of-a-predictive-model/.

Rowlands, Mark. 2000. *The Environmental Crisis: Understanding the Value of Nature*. New York: Palgrave Macmillan.

Rudin, Cynthia. 2019. "Stop Explaining Black Box Machine Learning Models for High Stakes Decisions and Use Interpretable Models Instead". *Nature Machine Intelligence* 1 (5): 206–15. https://doi.org/10.1038/s42256-019-0048-x.

Rudin, Cynthia, and Mit Sloan. 2013. "Predictive Policing: Using Machine Learning to Detect Patterns of Crime". *Wired*, August 22, 2013. https://www.wired.com/insights/2013/08/predictive-policing-using-machine-learning-to-detect-patterns-of-crime/.

Rudin, Cynthia, Caroline Wang, and Beau Coker. 2020. "The Age of Secrecy and Unfairness in Recidivism Prediction". *Harvard Data Science Review* 2 (1). https://doi.org/10.1162/99608f92.6ed64b30.

Russian Federation. 2017. "Examination of Various Dimensions of Emerging Technologies in the Area of Lethal Autonomous Weapons Systems, in the Context of the Objectives and Purposes of the Convention. Submitted by the Russian Federation". In *Item 6. Examination of Various Dimensions of Emerging Technologies in the Area of Lethal Autonomous Weapons Systems, in the Context of the Objective and Purposes of the Convention*. Geneva. https://admin.govexec.com/media/russia.pdf.

Ryan, N. J. 2018. "Five Kinds of Cyber Deterrence". *Philosophy & Technology* 31: 331–38. https://doi.org/10.1007/s13347-016-0251-1.

Salganik, Matthew J., Ian Lundberg, Alexander T. Kindel, Caitlin E. Ahearn, Khaled Al-Ghoneim, Abdullah Almaatouq, Drew M. Altschul, Jennie E. Brand, Nicole Bohme Carnegie, and Ryan James Compton. 2020. "Measuring the Predictability of Life Outcomes with a Scientific Mass Collaboration". *Proceedings of the National Academy of Sciences* 117 (15): 8398–403.

Samuel, Arthur L. 1960. "Some Moral and Technical Consequences of Automation—a Refutation". *Science* 132 (3429): 741–42. https://doi.org/10.1126/science.132.3429.741.

Sarantitis, George. 2020. "Data Shift in Machine Learning: What Is It and How to Detect It". *Georgios Sarantitis* (blog), April 16, 2020. https://gsarantitis.wordpress.com/2020/04/16/data-shift-in-machine-learning-what-is-it-and-how-to-detect-it/.

Sartorio, Carolina. 2007. "Causation and Responsibility". *Philosophy Compass* 2 (5): 749–65. https://doi.org/10.1111/j.1747-9991.2007.00097.x.

Savas, Onur, Lei Ding, Teresa Papaleo, and Ian McCulloh. 2020. "Adversarial Attacks and Countermeasures against ML Models in Army Multi-domain Operations". In *Artificial Intelligence and Machine Learning for Multi-domain Operations Applications II*, 11413:235–40. SPIE. https://doi.org/10.1117/12.2548798.

Schelling, Thomas C. 1966. *Arms and Influence.* New Haven: Yale University Press.

Schelling, Thomas C. 1980. *The Strategy of Conflict.* Cambridge, MA: Harvard University Press.

Scherrer, Nino, Olexa Bilaniuk, Yashas Annadani, Anirudh Goyal, Patrick Schwab, Bernhard Schölkopf, Michael C. Mozer, Yoshua Bengio, Stefan Bauer, and Nan Rosemary Ke. 2022. "Learning Neural Causal Models with Active Interventions". *arXiv:2109.02429 [Cs, Stat]*, March. http://arxiv.org/abs/2109.02429.

Schmitt, M. 2013. "Cyberspace and International Law: The Penumbral Mist of Uncertainty". *Harvard Law Review Forum* 126 (176): 176–80.

Schmitt, Michael N., ed. 2017. *Tallinn Manual 2.0 on the International Law Applicable to Cyber Operations: Prepared by the International Groups of Experts at the Invitation of the NATO Cooperative Cyber Defence Centre of Excellence.* 2nd ed. Cambridge: Cambridge University Press. https://doi.org/10.1017/9781316822524.

Schmitt, Michael N., and Jeffrey S. Thurnher. 2012. "Out of the Loop: Autonomous Weapon Systems and the Law of Armed Conflict". *Harvard National Security Journal* 4 (2): 231–81.

Schneier, Bruce. 2017. "Why the NSA Makes Us More Vulnerable to Cyberattacks". *Foreign Affairs*, May 30, 2017. https://www.foreignaffairs.com/articles/2017-05-30/why-nsa-makes-us-more-vulnerable-cyberattacks.

Schubert, Johan, Joel Brynielsson, Mattias Nilsson, and Peter Svenmarck. 2018. "Artificial Intelligence for Decision Support in Command and Control Systems". 15.

Schulzke, Marcus. 2013. "Autonomous Weapons and Distributed Responsibility". *Philosophy & Technology* 26 (2): 203–19. https://doi.org/10.1007/s13347-012-0089-0.

Schulzke, Marcus. 2016. "The Morality of Remote Warfare: Against the Asymmetry Objection to Remote Weaponry". *Political Studies* 64 (1): 90–105. https://doi.org/10.1111/1467-9248.12155.

Schwartz, Peter J., Daniel V. O'Neill, Meghan E. Bentz, Adam Brown, Brian S. Doyle, Olivia C. Liepa, Robert Lawrence, and Richard D. Hull. 2020.

"AI-Enabled Wargaming in the Military Decision Making Process". In *Artificial Intelligence and Machine Learning for Multi-domain Operations Applications II*, 11413:118–34. SPIE. https://doi.org/10.1117/12.2560494.

Sculley, D., Gary Holt, Daniel Golovin, Eugene Davydov, Todd Phillips, Dietmar Ebner, Vinay Chaudhary, Michael Young, Jean-François Crespo, and Dan Dennison. 2015. "Hidden Technical Debt in Machine Learning Systems". In *Advances in Neural Information Processing Systems*, vol. 28. Curran Associates. https://proceedings.neurips.cc/paper/2015/hash/86df7dcfd896fcaf2674f757a2463eba-Abstract.html.

Sechser, Todd S., Neil Narang, and Caitlin Talmadge. 2019. "Emerging Technologies and Strategic Stability in Peacetime, Crisis, and War". *Journal of Strategic Studies* 42 (6): 727–35. https://doi.org/10.1080/01402390.2019.1626725.

Select Committee on Artificial Intelligence. 2018. "AI in the UK: Ready, Willing and Able?" London: House of Lords.

Seppälä, Akseli, Teemu Birkstedt, and Matti Mäntymäki. 2021. "From Ethical AI Principles to Governed AI". In *2021 ICIS Proceedings*, 1–17. Austin, TX: Associaiton for Information Systems.

Sharif, Mahmood, Sruti Bhagavatula, Lujo Bauer, and Michael K. Reiter. 2016. "Accessorize to a Crime: Real and Stealthy Attacks on State-of-the-Art Face Recognition". In *Proceedings of the 2016 ACM SIGSAC Conference on Computer and Communications Security—CCS'16*, 1528–40. Vienna: Association for Computing Machinery. https://doi.org/10.1145/2976749.2978392.

Sharkey, Amanda. 2019. "Autonomous Weapons Systems, Killer Robots and Human Dignity". *Ethics and Information Technology* 21 (2): 75–87. https://doi.org/10.1007/s10676-018-9494-0.

Sharkey, Noel E. 2008. "Cassandra or False Prophet of Doom: AI Robots and War". *IEEE Intelligent Systems* 23 (4): 14–17.

Sharkey, Noel E. 2010. "Saying 'No!' to Lethal Autonomous Targeting". *Journal of Military Ethics* 9 (4): 369–83. https://doi.org/10.1080/15027570.2010.537903.

Sharkey, Noel E. 2012a. "Killing Made Easy: From Joysticks to Politics". In *Robot Ethics: The Ethical and Social Implications of Robotics*, edited by Patrick Lin, Keith Abney, and George Bekey, 111–28. Cambridge, MA: MIT Press.

Sharkey, Noel E. 2012b. "The Evitability of Autonomous Robot Warfare". *International Review of the Red Cross* 94 (886): 787–99. https://doi.org/10.1017/S1816383112000732.

Sharkey, Noel E. 2016. "Staying in the Loop: Human Supervisory Control of Weapons". In *Autonomous Weapons Systems: Law, Ethics, Policy*, edited by Claus Kreβ, Hin-Yan Liu, Nehal Bhuta, Robin Geiβ, and Susanne Beck, 23–38. Cambridge: Cambridge University Press. https://doi.org/10.1017/CBO9781316597873.002.

Shaw, Tyler, Adam Emfield, Andre Garcia, Ewart de Visser, Chris Miller, Raja Parasuraman, and Lisa Fern. 2010. "Evaluating the Benefits and Potential Costs of Automation Delegation for Supervisory Control of Multiple UAVs". *Proceedings of the Human Factors and Ergonomics Society Annual Meeting* 54 (19): 1498–1502. https://doi.org/10.1177/154193121005401930.

Shih, Andy, Arjun Sawhney, Jovana Kondic, Stefano Ermon, and Dorsa Sadigh. 2021. "On the Critical Role of Conventions in Adaptive Human-AI Collaboration". *arXiv:2104.02871 [Cs]*, April. http://arxiv.org/abs/2104.02871.

Shoemaker, David, ed. 2017. *Oxford Studies in Agency and Responsibility*. Vol. 4. New York: Oxford University Press.

Shue, Henry. 2008. "Concept Wars". *Survival* 50 (2): 185–92.

Simon-Michel, Jean Hugues. 2014. "Report of the 2014 Informal Meeting of Experts on Lethal Autonomous Weapons Systems (LAWS)". In *CCW/MSP/2014/3*. In *High Contracting Parties to the Geneva Convention at the United Nations*, Vol. 16, No. 2014, 1–5. https://undocs.org/pdf?symbol=en/ccw/msp/2014/3. Accessed October 2022.

Sinha, Aman, Hongseok Namkoong, and John Duchi. 2017. "Certifying Some Distributional Robustness with Principled Adversarial Training". *arXiv:1710.10571 [Cs, Stat]*, October. http://arxiv.org/abs/1710.10571.

Skerker, Michael, Duncan Purves, and Ryan Jenkins. 2020. "Autonomous Weapons Systems and the Moral Equality of Combatants". *Ethics and Information Technology* 22 (3): 197–209. https://doi.org/10.1007/s10676-020-09528-0.

Sparrow, Robert. 2007. "Killer Robots". *Journal of Applied Philosophy* 24 (1): 62–77. https://doi.org/10.1111/j.1468-5930.2007.00346.x.

Sparrow, Robert. 2016. "Robots and Respect: Assessing the Case against Autonomous Weapon Systems". *Ethics & International Affairs* 30 (1): 93–116. https://doi.org/10.1017/S0892679415000647.

Steinhoff, Uwe. 2013. "Killing Them Safely: Extreme Asymmetry and Its Discontents". In *Killing by Remote Control: The Ethics of an Unmanned Military*, edited by Bradley Jay Strawser, 179–208. Oxford: Oxford University Press. https://oxford.universitypressscholarship.com/view/10.1093/acprof:oso/9780199926121.001.0001/acprof-9780199926121-chapter-9.

Sterner, Eric. 2011. "Retaliatory Deterrence in Cyberspace". *Strategic Studies Quarterly* 5 (1): 65–80.

Stevens, Tim. 2012. "A Cyberwar of Ideas? Deterrence and Norms in Cyberspace". *Contemporary Security Policy* 33 (1): 148–70. https://doi.org/10.1080/13523260.2012.659597.

Stevens, Tim. 2020. "Knowledge in the Grey Zone: AI and Cybersecurity". *Digital War* 1 (1): 164–70. https://doi.org/10.1057/s42984-020-00007-w.

Stevenson, Ryan A., Joseph A. Mikels, and Thomas W. James. 2007. "Characterization of the Affective Norms for English Words by Discrete Emotional Categories". *Behavior Research Methods* 39 (4): 1020–24. https://doi.org/10.3758/BF03192999.

Stilgoe, Jack, Richard Owen, and Phil Macnaghten. 2013. "Developing a Framework for Responsible Innovation". *Research Policy* 42 (9): 1568–80. https://doi.org/10.1016/j.respol.2013.05.008.

Stoltz, Christopher. 2018. "Augmenting the AOR: EOD Airman Provides Critical Skillset to Army Forensics Team". Air Force. https://www.af.mil/News/Article-Display/Article/1581694/augmenting-the-aor-eod-airman-provides-critical-skillset-to-army-forensics-team/.

Stowers, Kimberly, Lisa L. Brady, Christopher MacLellan, Ryan Wohleber, and Eduardo Salas. 2021. "Improving Teamwork Competencies in Human-Machine

Teams: Perspectives From Team Science". *Frontiers in Psychology* 12 (May): 590290. https://doi.org/10.3389/fpsyg.2021.590290.

Strawser, Bradley J. 2013. "Introduction: The Moral Landscape of Unmanned Weapons". In *Killing by Remote Control: The Ethics of an Unmanned Military*, edited by Bradley Jay Strawser, 3–24. Oxford: Oxford University Press. https://doi.org/10.1093/acprof:oso/9780199926121.003.0001.

Strawson, Peter. 1962. "Freedom and Resentment". *Proceedings of the British Academy* 48: 1–25, Oxford: Oxford University Press.

Stumborg, Michael, and Becky Roh. 2021. "Dimensions of Autonomous Decision-Making". CNA. https://www.cna.org/CNA_files/PDF/Dimensions-of-Aut onomous-Decision-making.pdf?utm_source=Center+for+Security+and+ Emerging+Technology&utm_campaign=1280c55e66-EMAIL_CAMPAIGN_ 2022_01_27_02_11&utm_medium=email&utm_term=0_fcbacf8c3e-1280c55 e66-438318142.

Sweeney, Latanya. 2013. *"Discrimination in Online Ad Delivery"*. Rochester, NY: Social Science Research Network.

Switzerland. 2016. "Informal Working Paper Submitted by Switzerland: Towards a 'Compliance-Based' Approach to LAWS". March 30, 2016. In *Informal Meeting of Experts on Lethal Autonomous Weapons Systems*. Geneva. https://www.reach ingcriticalwill.org/images/documents/Disarmament-fora/ccw/2016/meeting-experts-laws/documents/Switzerland-compliance.pdf.

Szegedy, Christian, Wojciech Zaremba, Ilya Sutskever, Joan Bruna, Dumitru Erhan, Ian Goodfellow, and Rob Fergus. 2013. "Intriguing Properties of Neural Networks". *arXiv:1312.6199 [Cs]*, December. http://arxiv.org/abs/1312.6199.

Taddeo, Mariarosaria. 2012a. "An Analysis for a Just Cyber Warfare". In *Fourth International Conference of Cyber Conflict*, edited by C. Czosseck, R. Ottis, and K. Ziolkowski, 209–2018. NATO CCD COE and IEEE Publication.

Taddeo, Mariarosaria. 2012b. "Information Warfare: A Philosophical Perspective". *Philosophy and Technology* 25 (1): 105–20.

Taddeo, Mariarosaria. 2013. "Cyber Security and Individual Rights, Striking the Right Balance". *Philosophy & Technology* 26 (4): 353–56. https://doi.org/ 10.1007/s13347-013-0140-9.

Taddeo, Mariarosaria. 2014a. "Just Information Warfare". *Topoi*, April, 1–12. https://doi.org/10.1007/s11245-014-9245-8.

Taddeo, Mariarosaria. 2014b. "The Struggle Between Liberties and Authorities in the Information Age". *Science and Engineering Ethics*, September, 1–14. https://doi.org/10.1007/s11948-014-9586-0.

Taddeo, Mariarosaria. 2016a. "On the Risks of Relying on Analogies to Understand Cyber Conflicts". *Minds and Machines* 26 (4): 317–21. https://doi.org/10.1007/ s11023-016-9408-z.

Taddeo, Mariarosaria. 2016b. "The Moral Value of Information and Information Ethics". *In The Routledge Handbook of Philosophy of Information*, edited by Luciano Floridi, 90–105. New York: Routledge.

Taddeo, Mariarosaria. 2017a. "Cyber Conflicts and Political Power in Information Societies". *Minds and Machines* 27 (2): 265–68. https://doi.org/10.1007/s11 023-017-9436-3.

Taddeo, Mariarosaria. 2017b. "Deterrence by Norms to Stop Interstate Cyber Attacks". *Minds and Machines* 27 (3): 387–92. https://doi.org/10.1007/s11 023-017-9446-1.

Taddeo, Mariarosaria. 2017c. "Trusting Digital Technologies Correctly". *Minds and Machines* 27 (4): 565–68. https://doi.org/10.1007/s11023-017-9450-5.

Taddeo, Mariarosaria. 2018a. "How to Deter in Cyberspace". *European Centre of Excellence for Countering Hybrid Threats* 2018 (6): 1–10.

Taddeo, Mariarosaria. 2018b. "Deterrence and Norms to Foster Stability in Cyberspace". *Philosophy & Technology* 31 (3): 323–29. https://doi.org/10.1007/s13347-018-0328-0.

Taddeo, Mariarosaria. 2018c. "The Limits of Deterrence Theory in Cyberspace". *Philosophy & Technology* 31 (3): 339–55. https://doi.org/10.1007/s13 347-017-0290-2.

Taddeo, Mariarosaria. 2020. "The Ethical Governance of the Digital during and after the COVID-19 Pandemic". *Minds and Machines* 30 (2): 171–76. https:// doi.org/10.1007/s11023-020-09528-5.

Taddeo, Mariarosaria. 2022. "A Comparative Analysis of the Definitions of Autonomous Weapons Systems". *Science and Engineering Ethics* 28 (5): 37. https://doi.org/10.1007/s11948-022-00392-3.

Taddeo, Mariarosaria, and Alexander Blanchard. 2022. "Accepting Moral Responsibility for the Actions of Autonomous Weapons Systems—a Moral Gambit". *Philosophy & Technology* 35 (3): 78. https://doi.org/10.1007/s13 347-022-00571-x.

Taddeo, Mariarosaria, Alexander Blanchard, and Chris Thomas. 2023. "From AI Ethics Principles to Practices: A Teleological Methodology to Apply AI Ethics Principles in The Defence Domain". June 25, 2023. *SSRN Electronic Journal*. https://doi.org/10.2139/ssrn.4520945.

Taddeo, Mariarosaria, and Luciano Floridi. 2015. "The Debate on the Moral Responsibilities of Online Service Providers". *Science and Engineering Ethics*, November. https://doi.org/10.1007/s11948-015-9734-1.

Taddeo, Mariarosaria, and Luciano Floridi. 2018a. "Regulate Artificial Intelligence to Avert Cyber Arms Race". *Nature* 556 (7701): 296–98. https://doi.org/10.1038/ d41586-018-04602-6.

Taddeo, Mariarosaria, and Luciano Floridi. 2018b. "How AI Can Be a Force for Good": *Science* 361 (6404): 751–52. https://doi.org/10.1126/science.aat5991.

Taddeo, Mariarosaria, and Ludovica Glorioso, eds. 2016a. *Ethics and Policies for Cyber Operations*. New York: Springer.

Taddeo, Mariarosaria, and Ludovica Glorioso. 2016b. "Regulating Cyber Conflicts and Shaping Information Societies". In *Ethics and Policies for Cyber Operations*, edited by Mariarosaria Taddeo and Ludovica Glorioso, i–xvii. Berlin: Springer.

Taddeo, Mariarosaria, Tom McCutcheon, and Luciano Floridi. 2019. "Trusting Artificial Intelligence in Cybersecurity Is a Double-Edged Sword". *Nature Machine Intelligence* 1 (12): 557–60. https://doi.org/10.1038/s42 256-019-0109-1.

Taddeo, Mariarosaria, David McNeish, Alexander Blanchard, and Elizabeth Edgar. 2021. "Ethical Principles for Artificial Intelligence in National Defence".

Philosophy & Technology 34 (4): 1707–29. https://doi.org/10.1007/s13 347-021-00482-3.

Taddeo, Mariarosaria, Marta Ziosi, Andreas Tsamados, Luca Gilli, and Shalini Kurapati. 2022. "Artificial Intelligence for National Security: The Predictability Problem". London: Centre for Emerging Technology and Security.

Taleb, Nassim Nicholas. 2007. *The Black Swan: The Impact of the Highly Improbable*. New York: Random House.

Tamburrini, Guglielmo. 2016. "On Banning Autonomous Weapons Systems: From Deontological to Wide Consequentialist Reasons". In *Autonomous Weapons Systems: Law, Ethics, Policy*, edited by Bhuta Nehal, Susanne Beck, Robin Geiß, Hin-Yan Liu, and Caus Kreβ, 122–42. Cambridge: Cambridge University Press.

Tanji, Michael. 2009. "Deterring a Cyber Attack? Dream On . . .". Wired, February 19, 2009. https://www.wired.com/2009/02/deterring-a-cyb/.

Taylor, Isaac. 2020. "Who Is Responsible for Killer Robots? Autonomous Weapons, Group Agency, and the Military-Industrial Complex". *Journal of Applied Philosophy* 38: 320–34. https://doi.org/10.1111/japp.12469.

Terzis, Petros. 2020. "Onward for the Freedom of Others: Marching beyond the AI Ethics". In *Proceedings of the 2020 Conference on Fairness, Accountability, and Transparency*, 220–29. Barcelona: Association for Computing Machinery. https://doi.org/10.1145/3351095.3373152.

The Netherlands. 2017. *Examination of Various Dimensions of Emerging Technologies in the Area of Lethal Autonomous Weapons Systems, in the Context of the Objectives and Purposes of the Convention*. CCW/GGE.1/2017/WP.2. Group of Governmental Experts of the High Contracting Parties to the Convention on Prohibitions or Restrictions on the Use of Certain Conventional Weapons Which May Be Deemed to Be Excessively Injurious or to Have Indiscriminate Effects. Geneva: United Nations Office for Disarmament Affairs. https://www.reachingcriticalwill.org/images/documents/Disarmament-fora/ccw/2017/gge/documents/WP2.pdf.

"The UK and International Humanitarian Law 2018". n.d. https://www.gov.uk/government/publications/international-humanitarian-law-and-the-uk-government/uk-and-international-humanitarian-law-2018. Accessed November 1, 2020.

Tiwari, Sakshi. 2023. "Russia Threatens to Unleash 'Combat Robot' to Burn Ukraine's US & German-Origin Abrams & Leopard 2 Tanks". *Eurasian Times*, January 7, 2023. https://www.eurasiantimes.com/russia-threatens-to-unleash-combat-robot-to-burn-ukraines-us/.

Tobin, Donal. 2022. "What Is Data Cleansing and Why Does It Matter?" *Integrate. Io* (blog), January 21, 2022. https://www.integrate.io/blog/what-does-data-cleansing-entail-and-why-does-it-matter/.

Tossell, Chad, Boyoung Kim, Bianca Donadio, Ewart de Visser, Ryan Holec, and Elizabeth Phillips. 2020. "Appropriately Representing Military Tasks for Human-Machine Teaming Research". In Lecture Notes in Computer Science 12428, 245–65. https://doi.org/10.1007/978-3-030-59990-4_19.

Tsamados, Andreas, Nikita Aggarwal, Josh Cowls, Jessica Morley, Huw Roberts, Mariarosaria Taddeo, and Luciano Floridi. 2021. "The Ethics of Algorithms: Key Problems and Solutions". *AI & Society*, February. https://doi.org/10.1007/s00146-021-01154-8.

Tsamados, Andreas, Luciano Floridi, and Mariarosaria Taddeo. 2023. "The Cybersecurity Crisis of Artificial Intelligence: Unrestrained Adoption and Natural Language-Based Attacks". *SSRN Electronic Journal*. September 20, 2023. https://doi.org/10.2139/ssrn.4578165.

Tsamados, Andreas, and Mariarosaria Taddeo. 2023. "Human Control of Artificial Intelligent Systems: A Critical Review of Key Challenges and Approaches". July 9, 2023. *SSRN Electronic Journal*. https://doi.org/10.2139/ssrn.4504855.

Uesato, Jonathan, Brendan O'Donoghue, Aaron van den Oord, and Pushmeet Kohli. 2018. "Adversarial Risk and the Dangers of Evaluating against Weak Attacks". *arXiv:1802.05666 [Cs, Stat]*, February. http://arxiv.org/abs/1802.05666.

UK Government. 2014. "Deterrence in the Twenty-First Century: Government Response to the Committee's Eleventh Report". http://www.publications.par liament.uk/pa/cm201415/cmselect/cmdfence/525/52504.htm.

UK Government. 2015. "National Security Strategy 2016-2021". London: HM Government. https://www.gov.uk/government/uploads/system/uploads/atta chment_data/file/567242/national_cyber_security_strategy_2016.pdf.

UN GGE CCW. 2019. *Group of Governmental Experts on Emerging Technologies in the Area of Lethal Autonomous Weapons System, (2019). Report of the 2019 Session of the Group of Governmental Experts on Emerging Technologies in the Area of Lethal Autonomous Weapons Systems. Geneva: The United Nations Office at Geneva*. Geneva: United Nations Office at Geneva.

UN Institute for Disarmament Research. 2014. "Cyber Stability Seminar 2014: Preventing Cyber Conflict". https://unidir.org/wp-content/uploads/ 2023/05/cyber-stability-seminar-2014-en-612.pdf. Accessed July 2024.

UNIDIR (United Nations Institute for Disarmament Research). 2017. "The Weaponization of Increasingly Autonomous Technologies: Concerns, Characteristics and Definitional Approaches". UNIDIR Resources.

United Nations High Commissioner for Human Rights. 2014. *The Right to Privacy in the Digital Age: Annual Report of the United Nations High Commissioner for Human Rights and Reports of the Office of the High Commissioner and the Secretary-General*. A/HRC/27/37. Geneva: United Nations Human Rights Council.

United Nations High Commissioner for Human Rights. 2021. *The Right to Privacy in the Digital Age: Annual Report of the United Nations High Commissioner for Human Rights and Reports of the Office of the High Commissioner and the Secretary-General*. A/HRC/48/31. Geneva: United Nations Human Rights Council.

US Army. 2017. "Robotic and Autonomous Systems Strategy". https://www.tradoc. army.mil/Portals/14/Documents/RAS_Strategy.pdf.

US Department of Defense. 2012. "DoD Directive 3000.09 on Autonomy in Weapon Systems". https://www.esd.whs.mil/portals/54/documents/dd/issuan ces/dodd/300009p.pdf.

US Department of Defense. 2022a. "Fact Sheet on U.S. Security Assistance for Ukraine". May 10, 2022. https://www.defense.gov/News/Releases/Release/Arti cle/3027295/fact-sheet-on-us-security-assistance-for-ukraine/.

US Department of Defense. 2022b. "Responsible Artificial Intelligence Strategy and Implementation Pathway".

US Government. 2015. "The Department of Defense Cyber Strategy". Washington, DC.

US Senate Select Committee on Intelligence. 2002. *Joint Inquiry into Intelligence Community Activities before and After The Terrorist Attacks of September 11, 2001*. Washington, DC: U.S. Senate Select Committee on Intelligence and U.S. House Permanent Select Committee on Intelligence.

Veeramachaneni, Kalyan, Ignacio Arnaldo, Alfredo Cuesta-Infante, Vamsi Korrapati, Costas Bassias, and Ke Li. 2016. "AI2: Training a Big Data Machine to Defend". 13.

Verdiesen, Ilse, Filippo Santoni de Sio, and Virginia Dignum. 2021. "Accountability and Control over Autonomous Weapon Systems: A Framework for Comprehensive Human Oversight". *Minds and Machines* 31 (1): 137–63. https://doi.org/10.1007/s11023-020-09532-9.

Vogel, Kathleen M., Gwendolynne Reid, Christopher Kampe, and Paul Jones. 2021. "The Impact of AI on Intelligence Analysis: Tackling Issues of Collaboration, Algorithmic Transparency, Accountability, and Management". *Intelligence and National Security* 36 (6): 827–48.

Wagner, Markus. 2014. "The Dehumanization of International Humanitarian Law: Legal, Ethical, and Political Implications of Autonomous Weapon Systems". *Vanderbilt Journal of Transnational Law* 47: 1371–424.

Walch, Kathleen. 2020. "How AI Is Finding Patterns and Anomalies in Your Data". *Forbes*, May 10, 2020. https://www.forbes.com/sites/cognitiveworld/2020/05/10/finding-patterns-and-anomalies-in-your-data/.

Wallace, R. Jay. 1998. *Responsibility and the Moral Sentiments*. Cambridge, MA: Harvard Univ. Press.

Walliser, James C., Ewart J. de Visser, Eva Wiese, and Tyler H. Shaw. 2019. "Team Structure and Team Building Improve Human-Machine Teaming with Autonomous Agents". *Journal of Cognitive Engineering and Decision Making* 13 (4): 258–78. https://doi.org/10.1177/1555343419867563.

Walzer, Michael. 1977. *Just and Unjust Wars: A Moral Argument with Historical Illustrations*. New York: Basic Books.

Walzer, Michael. 2006. *Just and Unjust Wars: A Moral Argument with Historical Illustrations*. 4th ed. New York: Basic Books.

Watson, Gary. 1975. "Free Agency". *Journal of Philosophy* 72 (8): 205–20. https://doi.org/10.2307/2024703.

Weeramantry, C. G. 1985. "Nuclear Weaponry and Scientific Responsibility". *Journal of the Indian Law Institute* 27 (3): 351–86.

Weinbaum, Cortney, and John N. T. Shanahan. 2018. "Intelligence in a Data-Driven Age". *Joint Force Quarterly* 90: 4–9.

Whittlestone, Jess, Rune Nyrup, Anna Alexandrova, and Stephen Cave. 2019. "The Role and Limits of Principles in AI Ethics: Towards a Focus on Tensions". In *Proceedings of the 2019 AAAI/ACM Conference on AI, Ethics, and Society*, 195–200. Honolulu, HI: Association for Computing Machinery. https://doi.org/10.1145/3306618.3314289.

Widdershoven, Guy, Tineke Abma, and Bert Molewijk. 2009. "Empirical Ethics as Dialogical Practice". *Bioethics* 23 (4): 236–48. https://doi.org/10.1111/j.1467-8519.2009.01712.x.

Wiener, N. 1960. "Some Moral and Technical Consequences of Automation". *Science* 131 (3410): 1355–58. https://doi.org/10.1126/science.131.3410.1355.

Winter, Elliot. 2018. "Autonomous Weapons in Humanitarian Law: Understanding the Technology, Its Compliance with the Principle of Proportionality and the Role of Utilitarianism". *Groningen Journal of International Law* 6 (1): 183–202.

Winter, Elliot. 2020. "The Compatibility of Autonomous Weapons with the Principles of Distinction in the Law of Armed Conflict". *International & Comparative Law Quarterly* 69 (4): 845–76.

Wittgenstein, Ludwig. 2009. *Philosophical Investigations*. 4th ed. Translated by G. E. M. Anscombe. Malden, MA: Wiley-Blackwell.

Woodbury, Marsha. 2003. *Computer and Information Ethics*. Champaign, IL: Stipes.

Woods, D. D., E. S. Patterson, and E. M. Roth. 2002. "Can We Ever Escape from Data Overload? A Cognitive Systems Diagnosis". *Cognition, Technology & Work* 4 (1): 22–36. https://doi.org/10.1007/s101110200002.

Wooldridge, Michael J. 2020. *The Road to Conscious Machines: The Story of AI*. London: Pelican.

Wooldridge, Michael J., and Nicholas R. Jennings. 1995. "Intelligent Agents: Theory and Practice". *Knowledge Engineering Review* 10 (2): 115–52. https://doi.org/10.1017/S0269888900008122.

Yang, Guang-Zhong, Jim Bellingham, Pierre E. Dupont, Peer Fischer, Luciano Floridi, Robert Full, Neil Jacobstein, et al. 2018. "The Grand Challenges of Science Robotics". *Science Robotics* 3 (14): eaar7650. https://doi.org/10.1126/scirobotics.aar7650.

Yaron, Maya. 2018. "Statement by Maya Yaron to The Convention on Certain Conventional Weapons (CCW) GGE on Lethal Autonomous Weapons Systems (LAWS)". Geneva: Permanent Mission of Israel to the UN. https://www.unog.ch/80256EDD006B8954/(httpAssets)/990162020E17A5C9C12582720057E720/$file/2018_LAWS6b_Israel.pdf.

You, Sangseok, and Lionel Robert. 2016. "Emotional Attachment, Performance, and Viability in Teams Collaborating with Embodied Physical Action (EPA) Robots". https://aisel.aisnet.org/cgi/viewcontent.cgi?article=1810&context=jais.

Zagare, Frank C., and D. Marc Kilgour. 2000. *Perfect Deterrence*. Cambridge: Cambridge University Press.

Zegart, Amy B. 2005. "September 11 and the Adaptation Failure of U.S. Intelligence Agencies". *International Security* 29 (4): 78–111.

Zegart, Amy B. 2022. *Spies, Lies, and Algorithms: The History and Future of American Intelligence*. Princeton, NJ: Princeton University Press.

Zhuge, Jianwei, Thorsten Holz, Xinhui Han, Chengyu Song, and Wei Zou. 2007. "Collecting Autonomous Spreading Malware Using High-Interaction Honeypots". In *Information and Communications Security*, edited by Sihan Qing, Hideki Imai, and Guilin Wang, 438–51. Springer Berlin Heidelberg.

Index

www.ingramcontent.com/pod-product-compliance
Ingram Content Group UK Ltd.
Pitfield, Milton Keynes, MK11 3LW, UK
UKHW021909201225
466276UK00004B/10

9 780197 745441